THE MYSTERIES

of

SOUND AND NUMBER

BY

SHEIKH HABEEB AHMAD

"To everything there is a season, and a
time to every purpose under the Heaven."

KING SOLOMON

LONDON

~~NICHOLS & CO., 34 HART STREET, W.C.~~

~~AND~~ 23 OXFORD STREET, W.

1903

Kessinger Publishing's Rare Reprints
Thousands of Scarce and Hard-to-Find Books!

We kindly invite you to view our extensive catalog list at:
http://www.kessinger.net

PREFACE

—✦—

Two conclusions, among others, will probably be reached by thoughtful readers of this volume. The first will necessarily be that the very wonderful and mysterious law of Nature shown to be operative in connection with the rivalries of the turf does certainly exist. The second will be that such a discovery can only be the corner or fragment of some stupendous truth running through all human affairs.

It will not be surprising if the ultimate result of the discoveries now set forth, should be the development amongst us of a passionate desire for further knowledge concerning the manner in which unseen forces influence our daily lives. One can imagine, even, that many people among those who realise the full significance of the principles here enunciated, will be overcome by something like a feeling of alarm at the idea that their doings in everyday life are controlled, in a mysterious fashion, by Powers whose nature they cannot comprehend,—whose mere existence has hitherto been entirely unsuspected. They will wonder vaguely where such control stops short? To what extent are they the puppets of an invisible Will, or of forces with which they do not in the faintest degree know how to reckon? Others, again, more excited by the marvels of the law set forth than frightened at its possible range, will be keen to acquire further information concerning the way numerical values with which they may be associated (even if they are quite outside the pale of the special pursuit employed in this volume to illustrate the operation of the law) may bring them within the limits of possible calculations of a kind that would foreshadow impending events.

The following pages will be found closely and practically devoted to the proof of the law with which we are dealing; and for those who are chiefly interested in its application to the business of racing, the perusal of this Preface is quite unnecessary. But assuredly this book will be seriously considered by a large class of readers who will care more for the distant possibilities of knowledge to which it may lead, than for any immediate value, in a worldly sense, attaching to the instalment now conveyed to them.

We may confidently assume, in presence of the indications now given, that all great and important contributions to the knowledge of mankind are brought about under the guidance of unseen influences, certainly not confined in their operation to the affairs of the turf. Scientific discovery is allowed to move forward at some appointed rate. It looks like the outcome of this or that man's energy and genius. So the horse-race seems to be won by the superior effort or strength of the horse. And undeniably that effort, the health, temper, muscle, and so on of the horse, are factors in the result, just as the intellectual qualifications of the scientific discoverer are factors in his achievement; but the moral of the disclosure made in this volume is, that unseen influences are operative also to bring about results that would not happen without them. We are surrounded with unseen influences that are potent at every moment, and are subject to higher laws than those which have to do with chance or the caprices of human activity. And many circumstances of modern times will further point to the belief that these subtle laws are more directly associated with conscious intention on the part of some superhuman intelligence, than those which operate regularly through physical matter, and have specially engaged the attention of scientific observers during the nineteenth century. In the ages of faith most things that happened were credited to the interposition of Providence. An enlarged acquaintance with the laws of physics dissipated this idea in its early form, but now a still more enlarged experience is bringing back the original conception in a new and more reasonable shape. We do not have to reach in imagination to the level of the supreme Lord of the Universe in search of an explanation for every incident that seems to point to some sort of superhuman intervention; but we do recognise, or at all events many advancing thinkers are distinctly recognising, that the operation of the higher and till now more mysterious laws of the world are directed by specific intelligence and aimed at specific results. In the infinite complexities of Nature it is indeed impossible to say where blind law stops short, and overruling, intelligent Will of the higher order begins to work, but certainly intelligent Will *is* operative at some levels. And assuredly it is operative in connection with the expansion of human knowledge. Inquiring minds are illuminated, or eager investigation is defeated, according to whether on some higher levels of consciousness the time is regarded as ripe or not yet ripe for the introduction of new ideas into the current of the world's progress. And, furthermore, it may sometimes happen that experiments will be made by Higher Intelligence on the receptivity of the world at any given period. It does not follow by any means that merely because a great truth is enunciated it will be, to any considerable extent, appreciated. And if it is not appreciated in the right spirit, it is allowed to fall into neglect; no further efforts for the time are made along that line of development.

The present volume, in the estimation of the author and some of his friends, furnishes an example of such an experiment as that just described. For the first time in the history of the modern world a book has been written (which all men who choose to do so can read) in which a piece of knowledge hitherto kept back within a very narrow circle of peculiarly instructed pupils, pledged to secrecy, has been openly revealed. Many books have been written which declare the existence of "occult" knowledge, which set forth good reasons for believing that laws are in operation throughout the world that transcend the importance of those known to the physicist. But no occult secrets have been disclosed till now; no knowledge has been freely offered to the world at large which is calculated to invest those who assimilate it with *power* of any kind. But the light now cast upon some of the mysteries of Sound and Number do distinctly invest the student with a power of an entirely new kind,—the power, within certain limits, of divining future events of a certain order. Now, whoever realises the inner significance of such a power will see that it puts him in relation with realms of Nature, the very existence of which has till now been quite unsuspected by mankind at large,—suspected least of all by the most advanced races of mankind in the West. By what means is he to push forward investigation in reference to the body of law described? How is he to discover its further applications?

The answer that comes first is that real devotion to occult study will generally lead far—much further than the world at large imagines; but there is a second answer that may be more encouraging to people rather anxious to be helped than to take trouble. If the present instalment of occult secrets is properly appreciated, it is more than likely that further instalments will be forthcoming from the source from which these have been derived.

But what is meant here by proper appreciation? No very hard-and-fast rule can be laid down as to what will constitute its satisfactory reception, but it *may* be treated as merely of interest to those who are concerned with the pursuit selected to illustrate the working of the law described; or, on the other hand, it may come to be regarded as a deeply significant revelation of one way, at all events, in which the affairs of human life are controlled from higher planes of thought, activity, and consciousness. Thus if the present disclosures are generally received as important, even more for what they hint at than for what they set forth in plain language, then there is good reason to expect that further light will be thrown, in later works from the author's hand, on developments and ramifications of the principles now explained which will bring them into relation with other departments of human activity where new regions of practical usefulness may open out before them, and which more than ever will show the supremacy in human affairs of influences with which it is all important that we should put ourselves in harmonious relation.

TABLE OF CONTENTS

———◆———

THE
MYSTERIES OF SOUND AND NUMBER

———◆———

INTRODUCTION

VERY little attention has hitherto been paid by modern
students of the properties of numbers, to the vague records
transmitted to us from ancient Greece, of a belief enter-
tained by the philosophers of that period, to the effect that
numbers possessed some mysterious potency quite inde-
pendent of their arithmetical significance. The language
in which this belief is expressed is enigmatic for most of
us, in the highest degree. For example, let us take a few
sentences from Thomas Taylor's translation of Iamblicus'
Life of Pythagoras—

"The Pythagoreans received from the theology of
Orpheus the principles of intelligible and intellectual
numbers; they assigned them an abundant progression,
and extended their dominion as far as to sensibles themselves. Hence that proverb was peculiar to the Pythagoreans, that all things are assimilated to number.
Pythagoras, therefore, in *The Sacred Discourse*, clearly
says that 'number is the ruler of forms and ideas, and is
the cause of gods and demons.'"

Or let us take another example in which figures are
actually used, and we shall find the language just as
destitute of meaning for the modern mind—

"The tetrad was called by the Pythagoreans 'every
number,' because it comprehends in itself all the numbers
as far as to the decad, and the decad itself, for the sum

I

of 1, 2, 3, and 4, is 10. Hence both the decad and the tetrad were said by them to be every number, the decad indeed in energy but the tetrad in capacity. The sum likewise of these four numbers was said by them to constitute the tetractys in which all harmonic ratios are included. For 4 to 1, which is a quadruple ratio, forms the symphony bidiapason, the ratio 3 to 2 . . . etc., etc."

A very little information of that kind is more than modern European patience can stand, and the author of this volume wishes it to be clearly understood, from the outset, that he will not put any further strain of that nature on the endurance of the reader. He proposes, indeed, to deal with certain properties of numbers with which no arithmetic books have ever yet concerned themselves,—with some profoundly mysterious and deeply-seated laws of nature, to which the study of numbers in an entirely new way—new, that is, to the western world —will give a clue, and to put in plain language, for the service of all who read this book, the ancient teaching which underlay the ambiguous and intentionally unintelligible phraseology of the old Greek writers on the subject.

For, in truth, numbers have a deeper meaning than their mere arithmetical relationship would lead us to suppose. Their manipulation in a certain way will enable us to predict some events of a kind assigned by modern thinking to the arbitrament of chance. Throughout the ages in which the intellectual energies of the western world have been concentrated on the study of the physical universe, " Chance" has been allowed to usurp an ever-widening dominion in human affairs. Good luck or bad luck is accepted as the explanation of innumerable events which would be seen to fall within the range of specific causation, if our familiarity with natural law was deeper than it generally is. Curiously enough, a large body of metaphysicians contend that everything which happens, down to quite insignificant events, must be foreordained, because foreknown by Omniscience ; but the world at large, however respectful it may be in attitude towards the modern philosophers, treats their teaching, in practice, with supreme contempt, and goes on convinced that "it is better to be lucky than rich," and that throughout all our lives—from "the accident of birth" to the bad chance that brings on our last illness—we are victims of a lawless world, in which great care on our own part may enable us to escape serious disaster for a time, but in which "the chances are" that sooner or later we shall come to grief, like our neighbours.

Now, the studies of which this book is the fruit tend to eliminate the element of chance from life to a very considerable degree. The author does not aim at producing a metaphysical treatise on the providential government of the world. He will be concerned to a very limited degree only with abstract theorising on that subject, but he is in a position to explain some of the occult properties of numbers, in a way that will leave the fact that they *have* occult properties in doubt no longer, for anyone who has the good sense to peruse the following pages without flying off at a tangent, because the claim they make seems to him absurd on the face of it, before he has had time to realise that, absurd or not, it is substantiated.

It is difficult to forecast the scope of the present treatise without seeming to make arrogant claims in reference to the results it will accomplish, but the reader will find that it actually does disclose a system susceptible of being worked by any person of ordinary intelligence, by the application of which, to various problems in life, events may be predicted, even though they may seem to belong to the order of those which are regarded as the sport of Chance in a pre-eminent degree, or of rival activities in conflict with each other where no human sagacity can determine which will prevail.

No doubt, even this broad statement of the purpose the author has in view will excite the scornful ridicule of many critics, to begin with. That will become very much more scornful when they realise that proof of what has been said will be offered in connection with the practical investigation of chances that seem grotesquely beyond the reach of forecast. But the author may be forgiven for feeling some amusement at the thought that, while the most highly cultured of his readers will perhaps throw down the book with contempt as soon as they realise what it actually professes to accomplish,—secure in their certainty that " what they know not is not knowledge,"—others, on the contrary, will be less critical and more enterprising. They will try experiments, and they will find to their amazement that the rules now given to the world at large for the first time will enable anyone, willing to take a very moderate amount of trouble, to forecast, for instance, the result of a horse-race. There will be some practical charms connected with the acquisition of such a faculty, that will render the possessors indifferent to the question whether they are thought ridiculous or not. Supposing any considerable number of people do become possessed of such a faculty, it is one they will not be likely to hide " under a bushel." They will proceed to make use of it, and a widespread feeling of surprise will gradually ensue. Then perhaps the attention of people who study natural law for the sake of the knowledge to be gained will be turned in the direction of half-discovered mysteries, the importance of which is not to be measured by their apparent reasonableness,— by their harmony, that is to say, with knowledge of a more familiar kind.

" Half discovered " would perhaps be too flattering a phrase to apply to the discoveries set forth in this volume. The relationship of number with other natural phenomena is far too deeply seated to be completely interpreted by any wisdom but that which entirely transcends the possibilities of ordinary incarnate existence, and no attempt will be made in the following pages to explain why certain results ensue from the use of numbers, in accordance with principles now for the first time set forth in intelligible language. But it will be possible to show that such results are obtained. For these reasons the investigation on which the reader is invited to enter must not be conducted on lines corresponding with any familiar habits of scientific research. We cannot set out from broad conceptions of law, the main idea of which we understand, and from these deduce conclusions in reference to detail which seem logically to flow from the general principle. We must meekly follow out certain rules, however irrational they appear at the first glance, and ascertain whether they do, or not, give rise to important results. If they do, we shall manifestly be in touch with some profoundly obscure law of nature, the interest of which is not in the least degree impaired by the fact that its origin is lost in mystery. Nor must the significance of demonstrable facts be allowed to disappear, because it will be necessary to associate them with the phraseology of an ancient science now fallen into disrepute—Astrology.

The purpose of this book is not to argue for the authenticity of astrological theories which, as at present put before the western world, are, to say the least, fragmentary and probably distorted from the shape in which they were familiar to the ancient world when astrology was more seriously regarded. But if we blindly accept, as a provisional hypothesis, the fact that some hidden meaning lies in the numbers associated by ancient tradition with the various planets of the solar system, and make use of these numbers in ways that will be fully explained, we shall get at results that simply defy contempt or disregard. Further than this, we shall find that given numbers are associated with given sounds, and thus with given letters in the language of each country which are employed to indicate these sounds. Every student of language knows that the letters of the Hebrew alphabet were associated with definite numbers, but everyone does not know, what is equally true, that nature has associated number with all sounds, and therefore indirectly with the letters of every known alphabet, however little those who habitually use them may be aware of the fact.

The principle may seem perhaps in some degree the less unintelligible when we remember that every sound is a vibration of the air, and that such vibrations have a definite numerical value, as regarded from the ordinary scientific point of view. So many vibrations per second give a note of a definite pitch, and the thirds, fifths, and octaves of that note have vibrations in specific relation with the number of vibrations belonging to the fundamental note. At the first glance it is perfectly true that the numbers in question seem to have no relationship with those we get at in connection with the (occult) numerical value of sounds. That embarrassment, however, may be due to the fact that we have not yet been able to detect the relationship. It may lie hidden in the depths of the mystery connected with number, or it is possible that, if consonant sounds be examined with sufficient care, the numbers that occult science associates with them *will* be found to have some relationship with the vibrations per second to which they give rise. The inquiry is one on which we need not enter here, but when the very interesting results obtained by making use of the (hitherto) occult value of sounds shall be properly appreciated, future students will find a way to link the truths which this volume will establish with more familiar principles of physics, and it is not improbable that assistance in this direction may be derived from the use of that curiously delicate instrument of physical research, the Sensitive Flame. The flame is much more sensitive to the character of a sound than to its loudness. Every one who has used it, or seen it used, will be aware of its great sensitiveness to the hissing sound of the letter *s*, one of those which have a high value according to the law of numbers. Other consonant sounds have a higher value still, but cannot be made to continue in operation so long as the *s* sound, and it is thus more difficult to determine their exact effect on the flame; but a careful research might show that their momentary effect would be greater than a momentary enunciation of the *s*.

Meanwhile it will be readily perceived that if numbers have a relationship with letters, by virtue of the sounds which those letters represent, they must necessarily have relation also with names, which are merely so many sounds expressed by letters. Thus any given name can be represented by a number,—the aggregate or sum of the numbers associated with the letters that compose the name, or rather with the effective letters of it, those which are sounded as the name is pronounced. How do we get any forwarder, someone may ask, when we have assigned a number to a name? The name itself, it will be argued, is a purely arbitrary choice on somebody's part. It is a mere matter of " chance " what name, for example, parents may give to a child. They may waver about among many before they decide on any one. There can be no natural significance in a name! But at all events it must be recognised that a considerable school of thinkers have always maintained the " Determinist " position in philosophy—that nothing can happen without a cause, that even the seemingly spontaneous action of human creatures is the outcome of all the influences that have been brought to bear on them during life, and therefore might have been foreseen by perfect intelligence, a view of the matter that leaves no room for the favourite theory of chance. This book is not written to maintain one theory or the other as regards the mighty problem of Necessity and Free Will, but this reference to the Necessitarian view of the matter is merely put forward to remind the reader that there is nothing unheard of or absurd—if the adhesion of a great school of thinkers can guard a view from being absurd—in the idea that names may be given in accordance with the pressure of unseen influences that are all the more sure of their effect for being entirely unsuspected. That names and the numbers associated with names, according to the significance of the aggregate value of the sounds composing them, have a meaning quite outside and beyond what is commonly supposed, is one of the *truths*—not merely theories or beliefs—that this volume will demonstrate and establish. For some purposes the name of a person, a city, a country, or a horse may condense in itself a great deal of information concerning the destiny of the thing, place, or creature named. To read its significance we must, it is true, take other considerations into account concurrently. Time is a factor in the calculation, as will hereafter be shown, and that mysterious something which the modern world generally treats with contempt by reason of knowing nothing about it—planetary influence, must also be reckoned with. But planetary influences themselves are reducible to expression in number, and the calculations we have to make in dealing with names and their numerical significance are all arithmetical in their character.

There seems reason to believe that few of the events of human life are entirely unassociated with the occult influence of number; but, in order to prove the reality of such influence, it has been necessary to select some department of experience which will lend itself especially to the illustration of this mysterious truth. No impression will be made by the mere abstract enunciation of the theory. We must come to concrete examples of its operation in order to bring home conviction to the reader's mind. And the illustrations must be taken from a department of experience generally supposed to be quite outside the range of any successful forecast. Every one must be put into the position of being able to test the truth of the system to be described for himself, and the calculations required must be susceptible of being reduced to rule, and not too intricate. Under these circumstances the author has determined to show that "the glorious uncertainty" of horse-racing itself becomes amenable to the law of numbers, when these are properly applied to the investigation of the questions to which it gives rise. In connection with racing, fresh opportunities are offered every day for making trial of the numerical system of divination, and the truth of the law to be described will thus be more readily arrived at by the application of its formulæ to the records of the turf, than in any other way. Certainly, the purpose of this book is not to teach speculative enthusiasts how to back horses, with a reasonable likelihood of winning, but to establish the reality of hitherto unknown laws exerting a profound influence on human affairs. To accomplish this, the influence must be detected in operation, and it is most easily detected in cases where we have a number of named competitors engaged in some definite rivalry. Calculations of the same nature as those relating to the racecourse will apply to the competitive examination of a number of candidates for any given office or appointment, but this curious state of facts would be much more difficult of proof than the corresponding state of facts in reference to racing, which takes place somewhere every day under conditions which, as it will be seen, bring it very readily within range of the law of numbers.

If any one imagines that the main purpose of the author of this book has been to promulgate a method of winning money on the turf, let the transparent stupidity of such an idea be its own condemnation. Supposing the sordid motive to be in operation, it would best be served by strict secrecy in reference to the methods of calculation we are about to set forth. Their disclosure, by the time their trustworthiness has been generally appreciated, will tend to render them valueless for sordid purposes, for betting will be much more effectually discouraged when the events to which it relates are shown to be within the range of certain forecast than by any hostile legislation. Furthermore, the methods of calculation to be herein described will not be of a kind that will enable the student to foretell, a long time in advance, the result of a future race. But it will enable him to know what horse is going to win, *from the moment the race is started*; and the possibility of forecasting the result, even at this stage, will be quite enough to show that influences are in operation to control events generally supposed to be the sport of chance or the outcome of causes lying entirely on the physical plane, of a nature quite beyond the suspicion of the ordinary world at present. And this is the great truth, the establishment of which is really the main purpose of the present volume. In many ways the extreme materialism that characterised the thinking of civilised Europe during the latter half of the past century has been challenged and shaken. Multitudes have realised that the visible world is but one aspect of nature, and that an infinite variety of forces play upon it from unseen realms of activity. But greater multitudes still remain untouched by the subtle evidences of super-physical agencies at work all around us. The present undertaking is one that may perhaps appeal to them with good effect. Very often when students of what have been called "Nature's Mysteries" have been attempting to explain some of the loftier faculties that mankind may be invested with by means of psychic development, the question will rudely be put by scoffers, "Can you tell us what's going to win the Derby?" That inquiry has generally been treated by occult students with silent contempt, as evidencing such an imperfect comprehension of the purposes with which occult study is undertaken, as to make it hopeless to explain matters to any one who could propose such a test. But suppose it is for once taken seriously, and suppose the scoffers are not precisely told what is going to win the Derby, but enabled, subject to some restrictions that will not entirely preclude them even from making practical use of their knowledge, to find out for themselves what is going to win the Derby, or any other race in which they may be interested, what will be the result of such a demonstration? That remains to be seen; but anyhow the demonstration proposed is provided in the present volume, and it may have the effect of extending serious views of life over areas of humanity previously content to eat, drink, and be merry. Even in our sports we are in the midst of unseen power, control, and influence. Our own planning, our own efforts, are at best but factors in the sum-total of the causes at work, and the very efforts themselves, or the plans we set down to our spontaneous initiative, are shown to be the product, for ordinary mankind at all events, of unseen, unsuspected influences playing on our thoughts and inclinations. When that condition of things is fully realised along lines that admit of no mistake, inferences of wider significance still can hardly fail to suggest themselves in a natural sequence. So even for those who apply the Derby test to the claims of super-physical teaching, the answer they require is at last forthcoming ; while for more advanced inquirers, to whom the mysteries of practical occultism may hitherto have been matter rather for belief than for accurate comprehension, the present exposition of methods to be employed in analysing the subtle influences that control events of a certain class on this plane of life will be fraught with intense interest.

S. H. AHMAD.

CHAPTER I

THE NUMERICAL VALUE OF SOUNDS

EASTERN students of natural mystery are very generally aware of the existence among the Arabs of a certain science or method of divination which they call "Jaffar." The professors of this art or knowledge have a maxim, sometimes set forth in a metrical shape in its original tongue, the significance of which may be given in English as follows—

"It is good to be less than the opponent of one's own kind and to be greater than the opposite; but if both happen to be of the same number, the one who is younger will be the victor."

This utterance is hopelessly enigmatic for readers who simply encounter it in a philosophical treatise, without having any clue to its meaning beyond that which the words themselves may suggest. It is not likely that any unassisted student would be able to make sense of the sentence, and even those who would be inclined to believe that such utterances, when undeniably traceable to ancient sources, are always based upon some important truth, would still feel, in reference to the sentence just quoted, that the author, whoever he may have been, did not intend the maxim to be legible for all who passed. Such cryptic memoranda were only designed for the use of those who were already instructed in the truths thus half expressed and half concealed.

Some dozen years ago the present writer, always, from hereditary influence, on the alert to pick up any fragments of occult knowledge that might be available, found among family papers an old manuscript—some centuries old—which contained the above formula in Persian. He made many attempts to ascertain its meaning, and at last met a teacher who was able to throw some light on the subject and to explain that the rule had reference to determining the probabilities of success where several competitors were engaged in the pursuit of a common object. The explanation seemed to put a very ignoble face upon a mystery that had at first been regarded as probably carrying with it some lofty spiritual meaning. However, taking it at its humbler value, the author made a few experiments with it, the results of which were very striking, whether applied to problems arising from military history or modern litigation. At once, however, a difficulty arose. How could it happen that if two persons were engaged in frequent litigation, one would sometimes be successful and at another time the other? The numerical value of names, to be presently explained, was not alone enough to go upon, or certainly not in all cases. The rule was applied to lists of persons engaged in competitive examinations, again with curious success in so many cases as to rule out the hypothesis of chance, but it was only when the author fastened on the idea of testing the system by its application to horse-racing that the large field of experiment thus opened out led to the gradual development of the rules laid down in this book.

The necessity of combining influences having to do with time with those connected with numerical values was very soon apparent, but it was only by very slow degrees that the author realised the rapidity with which these influences change. It would tire the reader's patience if he merely described the course of his researches and experiments in pursuit of some practicable methods of working with the law embodied in the cryptic formula quoted above; but, having now indicated the nature of the clue which he originally obtained, it will be best to deal at once with the final results of his protracted study in such a way as may soonest enable the reader to check their significance by experiments of his own.

Before coming to close quarters with the methods of calculation to be presently explained, it is necessary to say a few words with reference to what may be called the natural and artificial values of figures. $1+2+3=6$ by any kind of reckoning. But 123, written so, carries, by ordinary convention, a meaning which far outruns the natural significance of the digits. The position of the figures by the decimal system of notation is of supreme importance, and the present writer has no wish to quarrel with the conventions of arithmetic when used for the purpose of facilitating calculations. But there are purposes to which figures can be applied, in connection with which it is necessary to value numerical expressions in a different way. Thus 123, for the purposes of the calculations with which this book will be chiefly concerned, must always be regarded as involving a significance expressed by the sum of the digits,—6 in this case. In the same way any numerical expression can be reduced to a single number by the addition of the digits, and the further addition of the digits expressing the sum if they are several in number. Thus 8165, taking an expression at random, becomes 20 by the process that may, for convenience, be called natural addition, and 20 becomes 2 by the same method. Or 74205 becomes 18 by natural addition, and the 18 becomes 9, which is therefore the inner significance, for certain purposes, of the expression 74205. The same result is reached in a different way if we divide any high number by 9, a figure which has many curious attributes, as the ordinary arithmetician is well aware, though generally content to amuse himself with them without going any more deeply into the subject. Thus 4126 divided by 9 gives 458 as a quotient with 4 as a remainder, while by natural addition the sum of the digits in the expression 4126 gives 13, and the sum of 13 gives 4. If we take an expression exactly divisible by 9, like 4122, there is no remainder, while the sum of the digits gives 9; but the two results are practically equivalent, for it is no less true to say that the expression in question is divisible by 9, with 9 as a remainder, than to say it is exactly divisible by 9. The sum of the digits in any expression exactly divisible by 9 will always be 9, when the process of natural addition is complete.

Just as any numerical expression, however long, can be reduced by natural addition to a single digit, so any name can be made to yield a single digit when the sum of the digits representing its gross numerical value is manipulated in the right way. The importance of ascertaining the numerical value of names will be explained later on. The gross numerical value is arrived at in each case by the addition in the ordinary way of the numbers representing the sounds embodied in the name, and then by the natural addition of the digits expressing that sum. A simple example will show the principle quite clearly. Take the name "John." The letters that spell that name will guide us to its numerical value, but this example will at once show that we must work with those letters only that represent sound, ignoring those which have no such value. The first sound expressed by the letter *j* has for its numerical value 3. The short *o* is valued at 2. The *h* is ignored, and the *n* is valued as 50. So the total value of the name is 55, or, by natural addition, 10, which again equals 1. Or, to take an example which happens to yield larger figures, "Smith," the *s* gives us 60, the *m* 40, the *i*, for reasons that will be hereafter explained, is not reckoned, the *t* and *h* give together 405, so that the gross numerical value of the name is 505, and that comes out again by natural addition, 1. The name "Johnson" would yield, besides the values already ascribed to the first syllable, 60 for *s* (*o* is not reckoned) and 50 for *n*, or in all, 165, which, by natural addition, gives 3 as the significant number of the name.

Now, to enable the reader to ascertain the numerical value of any name for himself, we proceed to give a complete list of the values of the sounds composing the

English language. Of course, the impatient reader will say at once there is no sense in the scale or progression adopted, no conceivable reason for assigning a high value to the letter or sound *t*, and a low value to the sound *d*. But the patient reader will wait to see whether, if he accepts this apparently incongruous classification of sounds and numbers, any practical results can be worked out that way. And for his comfort, in advance, it may just be worth while for him to remember that the numbers and the sounds were first discovered to have a relationship by students who worked with the Arabic language, in which the groupings of sounds and numerical values does not look quite so arbitrary.

NUMERICAL VALUE OF ENGLISH SOUNDS.

A as an open sound, as in the word "father"	1
B	2
J, *Ch*, or *G* soft, as in "George"	3
D	4
E in beginnings of words, or *H* as an aspirate in beginnings	5
W or *V* when ending a word, *O* long, as in "over," or *U*	6
Z	7
H (Arabic)	8
TT	9
E in the middle, or sounded at the end of a word, *I* and *Y*	10
K, *C* hard, *G* hard, or *Q*	20
L	30
M	40
N	50
S or *C* soft	60
O (Arabic)	70
F, *V*, beginning a word or syllable, *Ph* or *P*	80
SS (Arabic)	90
Q (Arabic)	100
R	200
Sh	300
T	400
X	110
Th, as in "the"	4
Th, as in "thing"	405

The sounds in the foregoing list which are indicated as "Arabic" cannot well be represented by the English letters used, but are inserted here partly in order to make the progressive series of numbers complete, and partly because the sounds in question may occur in some names of foreign origin. The Arabic *H* has a slightly guttural flavour, best indicated for the European ear by the concluding sound in the German word *hoch*. The Arabic *O* is equivalent to the Hebrew *ain*, but the sound does not occur in any European language. The Arabic double *SS* is sometimes transliterated *tz*, but those letters do not represent the real sound. It consists in an initial emphasis which slightly broadens the succeeding vowel. The English word "swan" illustrates it, and the sound *s* in that case would be counted 90.

The Arabic *Q* is equivalent to the Hebrew *koph*, but is hardly, if ever, represented in any European language.

This table must be supplemented by some qualifications before it can be practically used.

Vowels at the beginning of a name must in most cases be reckoned as worth 1 more than the simple value. Thus in the name of the town "Eton" the *e* would be counted as 11, not 10. But if it was hardly sounded at all, merely qualifying the pronunciation of the first consonant, as in "Edward," it would merely be counted 5. In the word "odour" the *o* is long, and at the same time a beginning vowel, so it would have 1 added to its value, and be counted 7. In the word "ox" the short *o* is counted as 2, and short *o*'s have the same value in the middle of words, as in "flower" or "morning."

When a still greater emphasis is laid on an initial *a* than is laid on the *e* in "Eton,"—as, for example, in the name "Amy,"—the *a* is treated as a diphthong, and has the value of *e*, or 10, added to it, the numerical sum of the name "Amy" being then 61 (the single significant number, the sum of 6 and 1, being 7). But when an initial *a* retains its open sound, as in the name "Arthur," it is not treated as a diphthong, but, on the other principle, has 1 added to its value. If there is very little emphasis on an initial *a*, and the sound of the word passes on at once to the first consonant, as in the name "Amelia," the *a* is only counted as 1, not even having an extra 1 added; but where a slightly different intonation has to be allowed for, as in the names "Allen" or "Anderson," then the *a* is treated as a diphthong, and counted 11.

Already it will be apparent that the task we have to perform in calculating the numerical value of a name is concerned entirely with sound, and only with spelling in so far as the spelling gives us a clue to the sound produced when the name is pronounced. We must never be too much influenced by spelling, but must think how we should write down a name phonetically if we had never seen it on paper and were simply endeavouring to choose letters that would accurately suggest the sound. The problem is, of course, embarrassing in a special degree when we are dealing with the English language. In Arabic and Persian, orthography is truly phonetic, and the determination of the numerical value of names is thus a simpler matter for students dealing with those languages; but English spelling is often a snare in the path of any one attempting to work with the methods under discussion, rather than a safe guide. As we go on with further qualifications of the original table, it will be seen that these are not required because of any irregularities in the natural law in operation, but simply because of the eccentricities of the language with which we are here concerned.

The vowel *i* in English is so little sounded in many words, that more often than not it may be ignored in counting up the value of a name. In the name "Martin," for instance, it would not be reckoned at all. But in the beginning of some names, as, for instance, in the case "India," it has some value, and would be counted. All depends on the part the vowel plays in producing the sound made when the name is pronounced. In the

name " London " there is no *o* sound produced when the
word is uttered.   For numerical purposes, therefore, we
simply count the consonants, omitting both *o*'s.   So with
the *u* in " Bucks " or the *i* in " Minster."   It is hardly
necessary to point out that such words as " cough,"
" plough," and " through " must be thought of as though
spelled *kof*, *plow*, and *thru* before we begin to count the
value of the sounds.

Double vowels, like the *o*'s in " Moon," are generally
to be taken as single.   Mute final *e*'s are of course ignored,
as in the name " Kate," but when the form " Katie " is
used the *i* and *e* together are equivalent to one *e*.

In the middle of words the vowel *u* often has the
sound of *e*, *u*, and the value of the *e* must be added
accordingly.

Coming now to the consonant sounds, the principle
of ignoring those which are not sounded has to be
observed as much as in the case of the vowels, and
double consonants are generally to be treated as though
they were single, simply because in general the duplica-
tion of the letter, determined by the etymology of the
word, does not really affect its sound as pronounced.
But there are cases where the doubled consonant gives
rise to a stress upon the sound, and then the numerical
value is altered.   A double *p* in this way, in the name
" Rapparee," counts 2, not 80, as the single *p* in the
word " rapacity " would be counted.   The rule may seem
very unintelligible, because at the first glance one would
expect the more emphasised *p* sound to have a higher
number attached to it than that which is less emphasised ;
but these values are derived from some obscure natural
laws, connected no doubt with the actual rate of vibration
which different sounds set up, and have been worked out
into cut-and-dried rules with reference to another language
than English.   Anyone who will take the trouble to
learn Arabic will find the arbitrary irregularities of the
system less conspicuous.

The sound of *x* in the middle or end of a name is

almost always equivalent to that of *k* and *s*, so the value is accordingly 20 + 60, or 80. In a few cases *x* at the beginning of a name pronounced like *z*, as in " Xenie," would have the value of *z*, that is 7.

Sometimes it may seem that a consonant is sounded when really the other letters present in the word would give the same sound without it. Thus the word " match " has merely the sound that could be equally expressed by the other letters if the *t* were omitted. In such a case it has no numerical value. The same remark applies to such a word as " luck," where the *c* has no effect on the sound.

Where the letters " tion," in words like " affection," yield the sound of *sh*, they are counted accordingly as though the word were spelled " affecshun." And the word " mansion " must be credited with the *sh* value.

Ch in " church " is, of course, counted for 3, and equally, of course, *ch* in Christian would be counted 20, as a *k*. But the German *ch* sound which sometimes makes its way into the English language, as in the Scotch word " loch," is counted as 600. The sound occurs in Arabic, in connection with which language it has been found to have that value.

In words including *ow* or *ou*, as in " fowl " or " soul," the two vowels together are equal to a long *o*, or 6.

Where *tw* occurs, as in " twin " or " twig," the value of *t* becomes 9 instead of 400, and the *w* is counted at its usual value, 6.

V is a very peculiar letter. When it occurs at the beginning of a name the value is generally 80 (the same value as *f*), but in the middle or at the end its value is generally that of the *w*, or 6. It is impossible to lay down a hard-and-fast rule that will cover all cases, and the student must be guided in this case, as in all others, by the actual sound produced as the word is pronounced.

Now, let us begin to consider the purposes to be served by ascertaining the numerical value of a name. Until the reader arrives at the proofs, which will be given

later on, of the profoundly mysterious law which operates
in connection with these values, the bare preliminary statement of the principle will seem wildly nonsensical, but so
would many of the familiar truths of science have seemed
nonsensical at an earlier stage of human development.
We are gradually approaching the revelation of a truth
belonging to the realm of occult science, and at the first
glance it will seem as hopelessly out of gear with any
familiar kind of knowledge, as, for example, any statement relating to the working of a telephone would have
seemed to the people, let us say, of the Elizabethan period.
But the discreet reader will not "shy" at it on that
account.   He will wait to see what evidence can be
brought forward in its support,—and that will be found
waiting for him in later pages of this volume.

The statement to which he is asked meanwhile, pro-
visionally, to listen, is this:—a name—whether it be
that of a human being, a horse, a town, or a book, is not
acquired by the person, animal, or place by accident.   It
is not the arbitrary choice of parent or owner, but
becomes the name of the person, animal, or place by
reason of unseen influences operating, quite unconsciously
to themselves, on the inclination,—which they imagine to
be quite free,—of the authorities who confer the name.
Something to that effect has been already said in the
Introduction, but the mystery involved has not yet been
fully explained.   The name given becomes a channel of
influence on the person or thing to which it belongs.   A
channel of what influence?—it will be asked, and the
answer merely seems to hurry us from one extravagant
theory to another still more incredible.   Indeed, we may
well shrink from any abstract statement of the mysterious
law at work, and leave the reader to frame the appropriate generalisation in any way he pleases when he once
understands what actually happens; but that which can
be shown by practical observation of events to take place
can at all events be stated in plain language, and the
search for an explanation as to why it happens so, can

be deferred until we realise the profoundly surprising truth.

At certain periods of each day certain influences, which for want of understanding their nature completely we must be content to call planetary influences, are predominant. The names that have been assigned to the days of the week are not mere arbitrary labels that have been affixed to them at random, but indicate quite correctly the influences that are broadly predominant during each day in turn. Thus Sunday is more associated with the influences which astrology attaches to the Sun than with those attaching to any other of the heavenly bodies. Monday is the Moon's day, in the same way. Tuesday is "ruled," in a certain sense, by Mars ; Wednesday, by Mercury ; Thursday, by Jupiter ; Friday, by Venus ; and Saturday, by Saturn. But subject to the undercurrent of influence, for example, of the Moon on a Monday, the successive *hours* of the day, reckoning from sunrise, are under the influence of the other heavenly bodies in a regular succession. Thus the first hour of a Monday is specially under the influence of the Moon, the second under that of Saturn, the third under that of Jupiter, and so on, in a manner to be described more fully hereafter. And the complexity of these "planetary" influences, as they are called in astrology, which treats both the Sun and Moon as planets for certain purposes, does not end here. Each hour is divided up into brief periods of four minutes each,—the time in which the Sun passes over one degree of longitude,—and each of these 4 - minute periods is "ruled" by one or other of the planets in regular succession.

Now, with each of the planets, from time immemorial, certain numbers have been associated. The number of Venus is 6, that of Jupiter 3, that of Saturn 8, and so on. Fuller details will follow later on, but for the moment our purpose is to show what all these apparently far-fetched explanations point to. When the reader understands what is the purpose of the calculations he is invited to attempt he will be better inclined to take the trouble to master the methods to be employed.

Any name, it has been shown, has a numerical value, and that value, however great, can be reduced by natural addition to a single figure. Any moment of the day falls within some one of the 4-minute periods above referred to. Now, whenever a number of persons, horses, yachts, or named things of any kind, are engaged in rivalry of any sort, whether that be a race, a competitive examination, a battle, or any other sort of struggle, the person or thing of which the name yields a number, corresponding with the number of the planet ruling the period within which the competition is decided, will win in that competition. Later on we may examine the question how far the coincidence of success and the numerical condition described is merely a subordinate circumstance, not implying that the planetary influence is the cause of the success. In the providential government of human affairs it may be that the success or failure of A, B, or C, in connection with any struggle in which they are engaged, is decreed with reference to conditions of justice, or the fitness of things, lying far deeper than the planetary conditions of the moment in which the struggle for success may culminate. But anyhow it will be found, by attentive observation, that the planetary influences, ascertained in the way described by the mysterious significance of numbers, do coincide with the result. And thus it happens that when we are dealing with struggles or competitions of a kind that must culminate within some definite planetary period, we can actually, if we know the names of the competitors, forecast, with something approaching certainty, the issue of the rivalry in question.

CHAPTER II

PLANETARY PERIODS

IN order that the reader may grasp the principle on which a practical use may be made of the calculations already described, it will now be necessary to consider the relation between number and time. Experience shows that certain numbers are, in some mysterious fashion, influential during certain periods of each day, and these numbers can most conveniently be identified by regarding them as associated with the planetary influences treated as succeeding each other in a certain order. Of course, there is more in this arrangement than a mere mnemonic device. The planetary influences are as real as the law of gravitation, though so very little understood at this stage of human knowledge generally; but, for the purpose of working the system we are in process of describing, it is not necessary to discuss the mystery of planetary influence. The names of the planets will merely be used to facilitate the needful calculations. And when we say that a given planet "rules" a given period, that for the present will merely be a short way of saying that during the period in question certain numbers have to be made use of in a certain way. So then, to begin with, the numbers attached (by a very ancient consent) to the heavenly bodies with which we are specially concerned are as follows :—

Saturn	.	.	.	8
Jupiter	.	.	.	3
Mars .	.	.	.	9
The Sun	.	.	.	4 and 1

Venus   6
Mercury   5
The Moon .   .   .   .   7 and 2

For the benefit of readers who may have paid some attention to astrological systems, a few words may usefully be said here with reference to the order in which the planets have just been enumerated. The order does not follow any obvious rule. First we get the three outer planets in a natural order. Then we go to the middle of the system, and then enumerate the inner planets in what most people will consider an unnatural order. But, firstly, the order given appears from experience to be the one in which the influences succeed one another during the day; and, secondly, if we count inwards from the outermost planet with which we are concerned for the first half of the series, and then outwards from the Sun for the second half, the only apparent error in that method has to do with the respective places in the solar system of Venus and Mercury. Now, a very strange state of things has to be noted here. In ancient Arabian books on astronomy the planet nearest the Sun is called Venus, and the next in order, going outwards, Mercury. At some periods, and for reasons that it seems difficult to make out, the later astronomers changed the names of the two inner planets, and called the one nearest the Sun Mercury, and the one nearest ourselves Venus. For astronomical purposes the names are mere labels identifying the planets in question, and it does not matter which way they are called. For astrological purposes the correct understanding of the truth is of supreme importance, because the influences associated with Mercury are quite different from the influences associated with Venus; and if we are right in declaring that the Mercurial influences attach really to the planet nearest to us, and the Venus influences to the one nearest the Sun, a great many calculations of modern astrologers are shown to be altogether wrong. The subject is not one on which it is necessary, for our present purpose, to spend a great deal of time, but anyhow, for the purposes of the calculations on which we are about to enter, we must treat the planetary influences as succeeding one another in the order given above.

And those influences operate in this way. Beginning at the moment,—the exact minute of sunrise at any given place,—the first hour (that is to say, the first sixty minutes after sunrise) is governed by the planet which rules the day of the week, whatever it may be. Thus, for example, Mars governs Tuesday as a whole, but in a special sense it governs the first hour. The Sun governs the second hour, Venus the third, Mercury the fourth, the Moon the fifth, Saturn the sixth, and Jupiter the seventh. But the hours themselves have to be divided into short periods of 4 minutes each, during each of which a different planetary influence prevails. Thus, for whatever hour we take, the planet which rules it as a whole governs the first 4-minute period. The next in order, the second, and so on. With a very little practice the progression is held in the mind so completely that no reference to any table is necessary to show what influence prevails at any given moment, but for beginners a table is necessary, and that may be constructed as follows, using the ordinary planetary symbols as found in any almanack. For convenience, we remind the reader that they are as follows :—

Saturn	.	.	.	. ♄
Jupiter	.	.	.	. ♃
Mars.	.	.	.	. ♂
The Sun	.	.	.	. ☉
Venus	.	.	.	. ♀
Mercury	.	.	.	. ☿
The Moon .	.	.	.	. ☽

The table shows the order in which the influences succeed one another, beginning always for each day with sunrise.

Hours	Sunday	Monday	Tuesday	Wednesday	Thursday	Friday	Saturday
1	☉	☽	♂	☿	♃	♀	♄
2	♀	♄	☉	☽	♂	☿	♃
3	☿	♃	♀	♄	☉	☽	♂
4	☽	♂	☿	♃	♀	♄	☉
5	♄	☉	☽	♂	☿	♃	♀
6	♃	♀	♄	☉	☽	♂	☿
7	♂	☿	♃	♀	♄	☉	☽
8	☉	☽	♂	☿	♃	♀	♄
9	♀	♄	☉	☽	♂	☿	♃
10	☿	♃	♀	♄	○	☽	♂
11	☽	♂	☿	♃	♀	♄	☉
12	♄	☉	☽	♂	☿	♃	♀
13	♃	♀	♄	☉	☽	♂	☿
14	♂	☿	♃	♀	♄	☉	☽
15	☉	☽	♂	☿	♃	♀	♄
16	♀	♄	☉	☽	♂	☿	♃
17	☿	♃	♀	♄	☉	☽	♂
18	☽	♂	☿	♃	♀	♄	☉
19	♄	☉	☽	♂	☿	♃	♀
20	♃	♀	♄	☉	☽	♂	☿
21	♂	☿	♃	♀	♄	☉	☽
22	☉	☽	♂	☿	♃	♀	♄
23	♀	♄	☉	☽	♂	☿	♃
24	☿	♃	♀	♄	☉	☽	♂

This table will serve as well to identify any given 4-minute period, with its appropriate influence, as to show the hour influence. For example, say we want to ascertain the operative influence for London, or some place near, at thirty-five minutes past two on Thursday, the 17th of July 1902. The almanack shows that, at Greenwich, the Sun rose on that day at four minutes past four. At noon, therefore, seven hours and fifty-six minutes had elapsed since its rising. At 2.35 so much more time had passed, or in all ten hours thirty-one minutes. We were therefore, at 2.35, under the influence of the planet ruling the eleventh hour of the day. As that was a Thursday, the planet of the eleventh hour was Venus. Thirty-one minutes of the hour had elapsed, so we were in the eighth 4-minute period. Turn to a Friday column, in which Venus governs the first term of the series, and it will be seen that she also governs the eighth term. So the Venus influence, or the influence of the number 6, was in operation at the moment taken for consideration.

In making any calculation of this nature having to do with time, the longitude of the place concerned must be taken into account if any use be made at that place of Greenwich time. If all the calculation has to do with local time, the calculation is as good for one place as another. At Exeter, for example, though that place lies three and a half degrees of longitude to the west of Greenwich, the Sun rose by local time at 4.4, just as in London. And if the question had been, what planetary influence was in operation at 2.35, by local time, on the day above dealt with, the answer would be the same for Exeter as for London. But places not so far away as Exeter have a trick of using Greenwich time, and then their position on the map must be taken into account. For example, at Rochester, only half a degree to the east of Greenwich, Greenwich time is probably used for most purposes, and then if we were asked what influence prevailed at Rochester on the day above named at 2.35, by Greenwich time, the easterly position must be taken into account.   At London, at the time mentioned, there was only one more minute left of the period governed by Venus.   But at Rochester, though the Sun rose by local time at the same moment as in London, by Greenwich time it really rose 2 minutes earlier.   And those 2 minutes would have shifted all the periods by a corresponding amount, so that at the moment above considered the Rochester people would have been one minute into the ninth period,—on that day the period governed by Mercury, with 5 as its significant number.   It will readily be seen that, in dealing with events passing so swiftly as those of the turf, the considerations just put forward may be very important.   They would just make all the difference between success and failure in forecasting a result.

Experience shows that the period influential in connection with a race is the period within which it is finished.   And as that period can only be forecast when we know the time at which the race is actually started, it is not possible by the methods of calculation herein described to settle days beforehand what horse will win any given race.   But from the moment the race is started the result can be foreseen, and, for practical purposes, it is possible beforehand to arrange for a series of contingencies, if we wish to be able, as soon as the race is started, to say what horse will win.   It does not often happen that the start is delayed for more than half an hour beyond the advertised time of starting, so if eight 4 - minute periods are provided for, we can generally forecast all the contingencies that can arise.   The time table would be drawn out as follows, for a time and date taken at random :—

Time, 1.40, Wednesday, May 28, 1902.   Place, London, or immediate vicinity.

Sunrise, 3.55   1 2

                        3.55

                        8. 5

                        1.40

                        ───

                        9.45

So we are in the twelfth period of the tenth hour, with three minutes to spare. We write down the numbers of the planets governing that and the few succeeding periods as follows :—

8						
6	5	7 and 2	8	3	9	4 and 1
1.43	47	51	55	59	2.3	2.7

We will suppose that such a time table as this has been prepared with reference to some race advertised to be run at the date and time mentioned. The next thing to do is to find out the numerical value of the names of the various horses engaged. For simplicity's sake, to illustrate the principle involved in a plainly intelligible manner, we will assume that 5 horses are engaged, and that the numbers of their names are as follows—A, 521 ; B, 654 ; C, 21 ; D, 1234 ; E, 995. Then, by natural addition, we get at these single numbers for each—A, 8 ; B, 6 ; C, 3 ; D, 1 ; E, 5. The reading of the calculation —subject to refined qualifications to be mentioned later —is as follows. If the race is started at such a time that it will be finished within the period ending 1.47, then E will win. If it is finished later, but before 1.51, a contingency arises of which we will speak directly. If the end is timed for the period between 51 and 55 minutes, A will win. For the next period C will be the winner. The period ending at 2.3 is to be specially provided for, as we will explain presently, and the winner, if the race is started so as to end between 2.3 and 2.7, will be D.

This example, let us hasten to add, will serve to give the reader a general idea of the system we are beginning to explain ; but various qualifications must be taken into account when it is actually applied to practical purposes. We only confuse the reader's mind if we endeavour to apply these all at once to the first illustrative examples. With the whole body of rules latent in his mind the student of the system will have no difficulty in applying the qualifications, but in describing them it is necessary to deal with these one at a time.

Let those who, reading so far, are inclined to laugh at such apparently absurd ideas wait till they see by actual records of past racing the extraordinary manner in which the result almost invariably comes out as described. Meanwhile, to complete the explanation of the rules to be applied to actual problems, let us deal with the contingencies in the above example, which are not provided for by the rule as so far given.

When there is no horse in the race whose number exactly corresponds with the planetary number of the winning period, nor with its "interchangeable" (the meaning of this expression will be described directly), the winner is found by the application of the very curious rule referred to at the beginning of Chapter I., concerning the lowest of its kind, and the greater of the opposite. That rule, as already explained, was really the nucleus round which the whole body of rules—the complicated system of numerical divination which this book sets forth —was actually framed. It will be much more readily understood by showing how it works in a concrete example than by means of any language that might express it in abstract terms.

When there is no horse in the race with a number corresponding to that of the winning period, we must take the lowest of the even numbers present and the lowest of the odd numbers, and then the winner will be the greater of those two. Thus, in the case with which we have been dealing, the lowest of the even numbers is 6; the lowest of the odd numbers, 1. 6 is greater than 1, therefore the horse B will be the winner if the race is run at such time as to finish in the periods ending at the 51st minute, or at 2.3.

This law of lowest and greater may under special circumstances be applied to the gross numerical values of names, instead of to the single digits deduced therefrom.

For example, if several horses in a race have the same

(condensed) number, and that corresponds to the winning period, their gross numerical values may be compared, and the lowest and greater rule applied. Or a very peculiar or special case may arise. If we find that the winning planetary period bears the same number as the hour number at the head of our time table, then the lowest and greater rule may be applied to the *gross* values, and if the name so indicated is reducible by natural addition to a number the same as those of the hour and planetary period, then that name may be taken as the winner. But it is obvious that this contingency is one that will not often arise.

But another contingency may often arise in practice which is not exemplified in the case just mentioned. Two or more horses in the race may have the same number. In that case—if that is the number of the winning period—the *youngest* of the similarly numbered horses will win. If two are of the same age, the horse with the name that has the largest opposite kind of gross value will win. By gross value, of course, we mean the sum of the values of the sounds in the name, 521 being supposed to be the gross value of the name of the horse called A in the above example.

One other consideration must be borne in mind in all calculations of this nature. Sometimes, when there is no horse with a number that corresponds with the planetary number of the winning period, it will suffice, without applying the rule just described about the lowest of each kind and the greater of the opposite, to work with an " interchangeable " number. The numbers of the Sun, 4 and 1, are interchangeable with that of Saturn, 8 ; also, in very rare cases, with those of the Moon, 7 and 2 ; but it is hardly necessary to confuse the present explanation with the contingency in question, which really has scarcely any bearing on the problems with which this book is concerned. The number of Mars, 9, is interchangeable with that of Mercury, 5, and that of Venus, 6, with that of Jupiter, 3. This law is

3

not a little embarrassing to students beginning such calculations as we have in hand, but it must not be left out
of sight, for in some cases the interchangeable number
has to be used in preference to that which may be called
the primary number. The point to be noticed will be
readily apprehended when we come to deal with practical
examples, but in order that this general explanation may
not remain in any way inaccurate, by ignoring important
qualifications of the rules laid down, let us add to the
above statement concerning interchangeable numbers this
caution. When the period in which the race is timed to
finish is a " negative " period, the interchangeable number
is to be taken as indicating the winner in preference to
the primary number belonging to that period.

What is the meaning of a negative period?

It is unnecessary at present to go into the meta-
physics of the subject, but everyone who has even brushed
the surface of metaphysical studies will be familiar with
the idea that the Positive and Negative principles—sometimes spoken of as the Male and Female principles—run
through nature in a great variety of ways. They have
to do with number, with the planets, and therefore with
such calculations as we are now concerned with. For our
present purposes it will be enough to define the meanings
of positive and negative periods of time.

~~The first series of seven hours after every sunrise is
"positive." The rest of the hours of that day up to
sunset are "negative." And within these limits the first
seven 4-minute periods of each hour are positive, and
the rest, up to the end of the hour, negative.~~

Now, when we are dealing—in such calculations as
are illustrated by the above example of an imaginary
race—with the planetary periods in which the competition may culminate, it is all important to consider whether
the winning period belongs to a positive or negative series
of 4-minute periods. It is true that the negative force
of a 4-minute period is somewhat weakened if it fall
within a positive hour, and *vice versâ*, but for practical

purposes the character of the short period is predominant.

If the winning (short) period for any race is a positive period, the horse with the corresponding number will win. If the winning period be negative, then the horse with the "interchangeable" number will win.

This is an all-important matter to remember, and without this keystone idea the whole system we are describing would fall to pieces and be found untrustworthy. The student attempting to work it with imperfect knowledge would fail, and would erroneously imagine that the law had broken down. The law never breaks down, but it is subtle in its operation, and takes every influence into account. If it ever seems to break down in his hands, the student may be absolutely sure that some factor has been omitted in his calculations. In this volume, at all events, there are no reservations, and if all its rules are absorbed into the student's mind and faithfully applied he will never find the law breaking down.

The explanations embodied in the last few pages must now be set forth in a brief form, to be thus the more readily remembered, though the brief statement would have had no meaning for the student without the foregoing explanations.

When the time table has been set out, mark periods as positive or negative, or take note of the state of facts concerning them, *i.e.* as to where the change takes place from the one character to the other.

For a positive period take, as the winner, the horse whose number corresponds with the number of that period.

If there is no horse with that number, take anyone which has the interchangeable number of the winning period.

If the period is negative, take the horse with its interchangeable number in preference to anyone with its own number.

Even if the period is negative, if there is no horse

with the interchangeable number, but if there is one with the primary number of the period, take that in preference to falling back on the lowest and greater rule.

It is now almost time to begin a series of practical illustrations of the way the law works, but the reader will so soon be wanting to try original experiments for himself, and it is so desirable that he should not make mistakes in setting out his time table, that although the method of doing this has been quite fully explained already, it seems worth while to give a few more illustrations in the nature of exercises for practice.

Wanted the time table for some place, on about the same meridian as London, for 1.30 p.m. on Saturday, the 4th January 1902. Time of sunrise, 8.8—

$$
\begin{array}{r}
1\,2 \\
8.8 \\
\hline
3.52 \\
1.30 \\
\hline
5.22
\end{array}
$$

The sixth hour on a Saturday is ruled by Mercury; the sixth 4-minute period of a Mercury hour is under the Sun, so the table stands :—

5					
4.1	6	5	7.2	8	3
1.32	36	40	44	48	52

Wanted a table for 3 o'clock, September 30, 1902— a Tuesday.

Answer—

6					
6	5	7.2	8	3	9
3.4	8	12	16	20	24

In this case, as the time of sunrise was exactly 6 a.m., the moment of 3 o'clock precisely completes the ninth hour, so we have to take the tenth hour in the first short period thereof for our table, with the whole 4 minutes assigned to the first term.

Wanted a table for 2 p.m., December 11, 1902—a Thursday.

Answer—

	8					
	8	3	9	4.1	6	5
	2.2	6	10	14	18	22

Let us now go on with practical tests of our great law by its application to some of the leading races actually run in the current year.

CHAPTER III

THE LAW ILLUSTRATED BY THE RECORDS OF HORSE-RACING

THE Derby of 1902 was run on the 4th of June, a Wednesday, at 3.19½, having, of course, been set for 3 o'clock. The length of the course was about one mile and a half. The sunrise on that morning was at 3.49. Our time table therefore would be made out as follows :—

$$
\begin{array}{c}
12 \\
3.49 \\
\hline
8.11 \\
3 \\
\hline
11.11
\end{array}
$$

Calculated for the advertised time, the race would thus be run in the twelfth hour after sunrise, which for a Wednesday (Mercury's day) would be ruled by Mars, of which the number is 9. The 11 minutes show us to have 1 minute left of the third period, which under the circumstances would be a Venus period, represented by the number 6. Our table therefore would be as follows :—

9							
6	5	7.2	8	3	9	4.1	6
1	5	9	13	17	21	25	29

The lowest column of figures, as the example already given will show, represents the number of minutes which may have elapsed after the advertised time for starting the race before the horses are actually " off."

The actual start was at 3.19½, and the time to allow for a 1½-mile course would be about 3 minutes, so it may be estimated (from the moment the actual start takes place) that the race would be finished at 22½ minutes past 3. The winning planetary period therefore would be last but one of our table governed by the Sun's numbers, 4 and 1.

Now we have to determine the numerical values of the names of the 18 horses in the race :—

Cheers.	Royal Ivy.
Fowling Piece.	Csardas.
Rising Glass.	Intruder.
Ard Patrick.	Robert le Diable.
Sceptre.	Duke of Westminster.
Friar Tuck.	Caro.
Prince Florizel.	Kearsage.
Royal Lancer.	Lancewood.
Pekin.	Water Wheel.

The numerical values of these names are calculated as follows :—

$$
\begin{array}{llll}
\text{Ch} & . & . & 3 \\
\text{e} & . & . & 10 \\
\text{e} & . & . & 0 \\
\text{r} & . & . & 200 \\
\text{s} & . & . & 60 \\
\hline
\end{array}
$$

273 = 3 by natural addition.

It will be seen that in accordance with rules already explained the double *e* is counted as only one.

$$
\begin{array}{lllll}
\text{F} & . & . & . & 80 \\
\text{o} & . & . & . & 6 \\
\text{w} & . & . & . & 0 \\
\text{l} & . & . & . & 30 \\
\hline
\end{array}
$$

Carry forward . 116

Brought forward	.	116
i .	.	.	.	.	0
n .	.	.	.	.	50
g .	.	.	.	.	20
				———
			186	= 6
				———
P .	.	.	.	.	80
i ⎫				
e ⎭ .	.	.	.	.	10
c .	.	.	.	.	60
e .	.	.	.	.	0
				———
			150	= 6
				——
			12	= 3

That is to say, the number of the double name as a whole would be 3, that of each component part 6.

R .	.	.	.	.	200
i .	.	.	.	.	0
s .	.	.	.	.	7
i .	.	.	.	.	0
n .	.	.	.	.	50
g .	.	.	.	.	20
				———
			277	= 7
				———
G .	.	.	.	.	20
l .	.	.	.	.	30
a .	.	.	.	.	1
s .	.	.	.	.	60
s .	.	.	.	.	0
				———
			111	= 3
				——
			10	= 1

A	.	.	.	.	2
r	.	.	.	.	200
d	.	.	.	.	4
					206 = 8
P	.	.	.	.	80
a	.	.	.	.	1
t	.	.	.	.	400
r	.	.	.	.	200
i	.	.	.	.	0
c, k	.	.	.	.	20
					701 = 8
					16 = 7
S	.	.	.	.	60
c	.	.	.	.	0
e	.	.	.	.	10
p	.	.	.	.	80
t	.	.	.	.	400
r, e	.	.	.	.	1
					551 = 2

At this point an explanation is necessary. It will cover a great many cases that arise in practice. The final *r e* of the name Sceptre cannot be reckoned as though they were sounded like the *r e* in the name " Trent," for example. There is a distinct tendency to roll the *r* in Trent. In the case of Sceptre the final syllable involves no such tendency. It is pronounced as though written Scepta, with a broad *a* at the end. We must therefore enumerate it accordingly, and do the same with most names ending in *e r*, like Westminster, to be presently noticed.

F	.	.	.	.	80
r	.	.	.	.	200
i	.	.	.	.	11
a, r	.	.	.	.	1
Carry forward				.	292 = 4

Brought forward   .   292 = 4

T	.	.	.	.	400
u	.	.	.	.	0
c ⎫					
k ⎭	.	.	.	20	

420 = 6

10 = 1

P	.	.	.	.	80
r	.	.	.	.	200
i	.	.	.	.	0
n	.	.	.	.	50
c	.	.	.	.	60
e	.	.	.	.	0

390 = 3

F	.	.	.	.	80
l	.	.	.	.	30
o	.	.	.	.	6
r	.	.	.	.	200
i	.	.	.	.	0
z	.	.	.	.	7
e	.	.	.	.	10
l	.	.	.	.	30

363 = 3

6 = 6

R	.	.	.	.	200
o	.	.	.	.	2
y	.	.	.	.	10
a	.	.	.	.	1
l	.	.	.	.	30

Carry forward   .   243 = 9

Brought forward . 243 = 9

L	.	.	.	.	30
a	.	.	.	.	1
n	.	.	.	.	50
c	.	.	.	.	60
e					
r	.	.	.	.	1

142 = 7

16 = 7

P	.	.	.	.	80
e	.	.	.	.	10
k	.	.	.	.	20
i	.	.	.	.	0
n	.	.	.	.	50

160 = 7

R	.	.	.	.	200
o	.	.	.	.	2
y	.	.	.	.	10
a	.	.	.	.	1
l	.	.	.	.	30

243 = 9

I	.	.	.	.	11
v	.	.	.	.	80
y	.	.	.	.	10

101 = 2

11 = 2

C ⎫
s ⎭ 60

a 1

r 200

d 4

a 1

s 60

—————

326 = 2

I 1

n 50

t 400

r 200

u 6

d 4

e ⎫
r ⎭ 1

—————

662 = 5

R 200

o 6

b 2

e 10

r 200

t 0

—————

418 = 4

—————

l ⎫
e ⎭ 0

—————

Carry forward   .   . = 4

Brought forward	.			. = 4	
D .	.	.	.	.	4
i .	.	.	.	.	10
a .	.	.	.	.	1
b .	.	.	.	.	2
l .	.	.	.	.	30
e .	.	.	.	.	0

$$47 = 2$$
$$—$$
$$6$$

In this case, as in that of all foreign names, the actual sound of the name as pronounced in the language to which it belongs has, of course, to be considered. The " le " may clearly be omitted as not constituting a part of the real name, in the same way that we would omit the article " the " in the English name " The Solicitor."

D .	.	.	.	.	4
u .	.	.	.	.	16
k .	.	.	.	.	20
e .	.	.	.	.	0

$$40 = 4$$

o ⎫ f ⎬ .	.	.	.	.	0

W .	.	.	.	.	6
e .	.	.	.	.	10
s .	.	.	.	.	60
t .	.	.	.	.	400
m .	.	.	.	.	40
i .	.	.	.	.	0
n .	.	.	.	.	50

Carry forward . 566—8

Brought forward   . 566—8

s   60
t   400
e ⎫
  ⎬   I
r ⎭
                        ————
                        1027 = 1
                             ——
                             5

In the name just enumerated the *u* in Duke is pronounced as if it were written *eu*, so we give it the value
of the two vowels, 10 and 6, or 16. The final *er* of
"Westminster" is valued as a broad *a*, like the *re* in
Sceptre.

C   20
a   I
r   200
o   6
                        ————
                        227 = 2

K   20
e   10
a   0
r   200
s   60
a   I
g   3
e   0
                        ————
                        294 = 6

L   30
a   I
n   50
c   60
e   0
w   6
o   0
o   0
d   4
                        ————
                        151 = 7

W	.	.	.	.	6
a	.	.	.	.	1
t	.	.	.	.	400
e					
r	.	.	.	.	1

$$408 = 3$$

W	.	.	.	.	6
h	.	.	.	.	0
e	.	.	.	.	10
e	.	.	.	.	0
l	.	.	.	.	30

$$46 = 1$$

$$4$$

We have now to determine which horse will win provided the race is started at the advertised time, and which will win if the start is delayed so as to bring the conclusion of the race within any of the later periods in our time table.

As 3 minutes (about) must be allowed· for the duration of the race, it is hardly worth while to consider the first period. The race would have to be started at 3 o'clock, to the moment, to render that operative. But say it started not later than 3 minutes past 3 (a very improbable contingency). Then we have to consider which horse would be the winner for the Mercury period.

It will be convenient to set out the final numbers in a way that renders comparison easy, as follows :—

Cheers	.	.	.	.	$273 = 3$
Fowling Piece	.	.	.	.	$336 = 3 — 6$
Rising Glass	.	.	.	.	$388 = 1 — 3$
Ard Patrick	.	.	.	.	$907 = 7 — 8$
Sceptre	.	.	.	.	$551 = 2$

Friar Tuck . . . . $712 = 1$—6
Prince Florizel . . . $753 = 6$—3
Royal Lancer . . . . $385 = 7$—7
Pekin . . . . . $160 = 7$
Royal Ivy . . . . $344 = 2$—2
Csardas . . . . . $326 = 2$
Intruder . . . . . $662 = 5$
Robert le Diable . . . $465 = 6$—2
Duke of Westminster . . $1067 = 5$—1
Caro . . . . . $227 = 2$
Kearsage . . . . $294 = 6$
Lancewood . . . . $151 = 7$
Water Wheel . . . . $454 = 4$—1

This table still requires some explanation. Where the name is double, as in the case of Fowling Piece, the numerical value of the most significant part of the name has to be considered in the choice of the winner. In the table the first column of single figures, as will be seen, represents the reduction (by natural addition) of the gross value of the double name; the second column the reduction of the significant part of the name where it is double, so we now have to compare significant part numbers with the numbers of single names. Intruder is the only 5 in the series, and would therefore be the winner for the Mercury period. Suppose there had been no 5 at all in the series, we should have fallen back on the lowest and greater rule, the operation of which in the case imagined may as well be at once explained. The lowest of the odds is 1, the lowest of the evens 2. There are five 2's in the company, and thus (all being of the same age) we have to consider gross values. It is important to remark that though all the horses in such a race as the Derby are nominally 3-year-olds, there must be differences of months in their ages, and these may not be easily ascertained. So the rule of the lowest and greater, where that has to be made use of, is not so entirely reliable as the direct numerical rules where they can be applied; at the same time, it does generally work even in cases where the exact age is uncertain. In other cases the lowest and greater rule is very trustworthy.

Coming back to the problem before us, we have discarded the single 1, and the choice now rests between the four 2's. Their gross values are as follows :—

Sceptre	551
Royal Ivy	344
Csardas	326
Robert le Diable . . .	465
Caro	227

Again, amongst these we have to apply the odd and even comparison. The lowest of the evens is Csardas, the lowest of the odds Caro. Csardas is greater than Caro, so he would have been the winner for the Mercury period had there been no Intruder.

Now for the Moon period. At the first glance we have two numbers of which to go in search—7 and 2. But we are enabled to discard the 7 because of the rule to be now described. There are 6 horses with Moon numbers :—

Pekin	7
Lancewood	7

and the five 2's already mentioned. The 7's may be left out of account because of the presence of the lowest odd number 1. This will seem a little confusing at first; but, it must be remembered, the lowest and greater rule is profoundly fundamental, and comes into play in one way or another whenever plain and exact numerical values do not give a direct answer to our questions.

As amongst the five 2's, the calculation for the Mercury period applies over again, and thus for the Moon period Csardas must be put down as the winner.

Now for the next or the Saturn period, we have one 8 to consider, and also (under the interchangeable rule) one 1—Water Wheel. We discard the 1 because the period

4

we are considering occurs in the positive revolution of the planetary periods. See the explanation given towards the end of Chapter II.

The Mars hour is the second Mars hour of the day, and consequently in the negative series; but the 4-minute period we are dealing with—the Saturn period of our table—comes in the first or positive series of the hour, therefore we work with the original number of the period, and not with its interchangeable. If the period had fallen within a negative series we should have worked preferably with the interchangeable number.

Ard Patrick is the only 8 in the company, and we therefore select him as the winner under the Saturn period.

We come next to the Jupiter period. There are three 3's to be considered. Being still in the positive portion of the hour, we need not take account of the interchangeable numbers. Of these the lowest of the odds is Cheers, and the one even is Rising Glass, and of these two the greater is Rising Glass, which would thus be the winner for the Jupiter period.

For the Mars period we have no 9, nor even its interchangeable number, 5. The lowest and greater rule must therefore be applied, with the result that Csardas is pointed to as the winner.

For the Sun period, we are now in the negative revolution of the 4-minute period, so we have to take the interchangeable 8, in preference to the only positive Sun number present, 1. Thus Ard Patrick becomes our selection for the Sun period.

We have still one more period to calculate—that of Venus. It is a negative period, the interchangeable number 3 is used in preference to the positive 6, and the same calculation adopted for the Jupiter period gives us Rising Glass as the winner.

Let us now set out these results in a manner which renders reference easy.

If the race is started at such a time as (allowing for

the 3 minutes, about, that it will take) will bring its conclusion within the—

Mercury period	.	.	.	Intruder wins.
Moon "	.	.	.	Csardas "
Saturn "	.	.	.	Ard Patrick "
Jupiter "	.	.	.	Rising Glass "
Mars "	.	.	.	Csardas "
Sun "	.	.	.	Ard Patrick "
Venus "	.	.	.	Rising Glass "

As a matter of fact, the race was actually started at 3.19½. Adding 3 minutes for the duration of the race, we have to look for the winner in the period including 3.22½. That is the Sun period, and Ard Patrick is the absolute winner.

But the above calculations tell us even more than the name of the absolute winner. We can determine with approximate certainty the names of the horses that will be " placed " second and third. These are indicated in more ways than one. To begin with, if there have been several horses with numbers that correspond to the winning planetary period, but if we have selected one among them as the winner—under any of the rules already described, then the placed horses will be found among the others of that same number. If there are no others of that number the numbers of the planetary period adjacent to the winning period must be taken into account, and in doing this we must work with the whole values of double names. Rules of this sort may appear very arbitrary, but they have been deduced from long experience of actual events.

In the case before us there are two horses with Sun numbers in the column of whole-name values— Rising Glass and Friar Tuck. These were the horses that were second and third. Other illustrations will be concerned with cases in which adjacent periods are effective.

We will now take another conspicuous race, and work out the problem it presents in the same way as in the above illustration.

The race for the Steward's Cup at Goodwood for the current year, 1902, was run on Tuesday, the 29th of July, and advertised for 2.45 p.m. The field was a very large one, and the list of horses started is given below, together with the numerical values of their names. It seems hardly necessary in this case to set out at length the manner in which the numerical values have been arrived at. The principles have been fully illustrated in the previous example, and the student can check his own calculations in all cases by reference to the appendix.

				Age.
Sundridge	317	2		4
Master Willie	747	9	1	6
Lord Bobs	302	5	3	4
Le Blizon	140	5	1	6
Water Shed	921	3		4
O'Donovan Rossa	456	6		5
Mauvezin	194	5		6
Lavengro	397	1		3
Bridge	205	7		6
St. Quintin	1040	5	7	3
Game Chick	93	3		3
Loch Doon	112	4	4	4
Noonday	120	3		4
Olympian	217	1		4
Blue Peter	529	7	5	3
Engineer	314	8		4
Portcullis	796	4		4
Mimicry	310	4		3
Royal River	523	1	2	4
Lady Macdonald	195	6		3
Shell Martin	1031	5	7	3
Aggressor	501	6		3

The time table, calculated for 2.45, will be—

	5						
6	5	7.2	8	3	9	4.1
47	51	55	59	3.3	7	11

In this illustration it will not be necessary to give a series of hypothetical winners for every possible period of starting. It will be enough to show the working of our law for the actual time of the race. The start was effected at 3 minutes past 3, and the assumed duration 1½ minute, so that the conclusion of the race is thrown into the Mars period, from the indications of which we must seek the winner.

There is no horse in the race, large as the field is, with the actual Mars number,—for though the whole name of Master Willie yields 9, we have already explained that in looking for actual winners we must always take the number of the significant part of the name.

We need not, in this case, consider the question whether the winning period is a positive or negative period, for in either case, with the actual number absent, we must take the interchangeable number. (It is just worth while to remark that as the winning period in this case is a negative period, we should have had to take the interchangeable number, even if a 9 had been present.)

The interchangeable number of 9 is 5. We have only one 5 in the company, and Mauvezin is at once identified as the winner.

The placed horses were Master Willie, with 9 (as the numerical value of the whole name), and O'Donovan Rossa, 6, the number of the adjacent planetary period.

With races in which only a few horses are concerned the calculations are naturally easier. Let us take, for example, the first race on Friday, August 1, at Goodwood. The set time was 1 o'clock. The horses running, all 2-year-olds, were—

| Fisher King | . | . | . | . | 670 | 4 | 9 |
| Castle Dance | . | . | . | . | 226 | 1 | 7 |

St. Asaph	.	.	.	.	692	8	1
Quaker's Wife		.	.	.	393	6	5
Blowing Stone		.	.	.	624	3	3
Night and Day		.	.	.	519	6	9
Maid of Clwyd		.	.	.	195	6	9

Sunrise for the day was 4.23, and the time table will thus be—

5			
8	3	9	4.1
3	7	11	15

The actual start was at 1.5, and the length of the course 5 furlongs—the assumed time for which would be $1\frac{1}{4}$ minute. The Jupiter period will therefore show us the result. The period is negative, but though the interchangeable number for 3 would be 6, there is no 6 present. We therefore accept the 3 which is present in the case of Blowing Stone, and point to him as the winner.

The second race of the same day was set for 1.30. The horses, 2-year-olds, running were—

Jennico .	.	.	.	.	89	8	
Quintessence .		.	.	.	660	3	
San Terenzo .		.	.	.	794	2	8
The Wag	.	.	.	.	41	5	9
Zaza .	.	.	.	.	16	7	

The time table was—

7.2			
8	3	9	4.1
31	35	39	43

The race was off at 1.31, the course 6 furlongs $= 1\frac{1}{2}$ minute. The Jupiter period indicates Quintessence as the absolute winner, with Jennico second.

CHAPTER IV

PRACTICAL DEMONSTRATIONS SHOWING THE WORKING OF THE LAW

THIS chapter simply deals with practical demonstrations showing, in an abbreviated form, the working of the law, for which purpose we have selected the last Epsom and Ascot weeks as being the most important.

EPSOM—*Wednesday*, the *4th June* 1902 (Sun rises, 3.48).

1st race, 1.30 p.m.

> Distance, 6 furlongs.
> Time occupied in running, $1\frac{1}{2}$ minute.
> " Off," $1.37\frac{1}{2}$.
> Finishes 1.39 p.m., under " 5."

Time Table.

8			
(4.1)	6	5	(7.2)
1.32	1.36	1.40	1.44

The first three horses—

Russet Brown	.	9 1 8 = 9	6 years.
Inishfree	.	. 6 4 1 = 2	5 ,,
Fosco .	.	. 1 6 8 = 6	Aged.

Now, under the law of interchange already explained, 9 wins in 5's time; 2 and 6 (one from the right and the other from the left adjacent columns) are placed second and third.

2nd race, 2.5 p.m.

    Distance, 5 furlongs.
    Time occupied in running, $1\frac{1}{4}$ minute.
    " Off," $2.13\frac{1}{2}$.
    Finishes $2.14\frac{3}{4}$ p.m., under " 8."

Time Table.

	3		
5	(7.2)	8	3
2.8	2.12	2.16	2.20

The first three horses—

Swift Cure	.	604 = 1	2 years old.
Doremi .	.	264 = 3	,,
Donzelle .	.	107 = 8	,,

The same law of interchange applies in this case also—
1 wins, 3 and 8 are placed.

————

3rd race is the " Derby Stakes," which has already
been explained in full details.

————

4th race, 3.45 p.m.

    Distance, 5 furlongs.
    Time occupied in running, $1\frac{1}{4}$ minute.
    " Off," $4.6\frac{1}{2}$.
    Finishes $4.7\frac{3}{4}$ p.m., under " 8."

Time Table.

9	(4.1)				
9	(4.1)	6	5	(7.2)	8
3.48	3.52	3.56	4.0	4.4	4.8

The first three horses—

Gun Club	.	.	.	$122 = \frac{5}{7}$	†
Fledgling	.	.	.	223 = 7	†
Pitch Dark	.	.	.	308 = 2	

The race ended in a dead-heat between the first two, about 4 minutes ahead of where it should have fallen in the ordinary course; but it must be remembered that Gun Club and Fledgling are the "compared" numbers, both are recently named or unnamed 2-year-olds. However, the most remarkable point in the event is, that all the three belong to one and the same planetary influence, namely, the Moon's.

———

5th race, 4.25 p.m.

    Distance, 5 furlongs.
    Time occupied in running, about 1 minute.
    " Off," 4.39.
    Finishes 4.40 p.m., under " 3."

Time Table.

	4				
	5	(7.2)	8	3	9
	4.28	4.32	4.36	4.0	4.4

The first three horses—

Miss Unicorn	.	$448 = \frac{7}{6}$	6 years old.
Princess of Ayr.		$752 = \frac{5}{1}$	3 „
Le Buff	. .	$122 = \frac{5}{1}$	Aged.

This case hardly requires any explanation. Under the law of interchange the race goes to " 6," 5's being placed.

———

6th race, 5 p.m.

    Distance, 1 mile.
    Time occupied in running, about 2 minutes.
    " Off," 5.11.
    Finishes about 5.13 p.m., under " 4.1."

Time Table.

	6			
8	3	9	4.1	6
5.4	5.8	5.12	5.16	5.20

The first three horses—

Flavus	.	.	$251 = 8$	Aged.
Vatel	.	.	$521 = 8$	,,
Clarissa	.	.	$302 = 5$	3 years old.

Nearly the same remarks apply in this case as in the preceding race.

————

Thursday, the 5*th June* 1902 (Sun rises, 3.47).

1*st race*, 1.30 p.m.

Distance, 5 furlongs.
Time occupied in running, about 1 minute.
" Off," 1.30$\frac{1}{2}$.
Finishes 1.31$\frac{1}{2}$ p.m., under " 8."

Time Table.

(4.1)
(7.2)   8   3
1.31   1.35   1.39

The first three horses—

Mixed Powder	243	$9 = 1$	2 years old.
Gourgaud	. 252	$= 9$	,,
Swift Cure	. 604	$= 1$	,,

In this case, though both Mixed Powder and Swift Cure are described as 2-year-olds, Mixed Powder was probably the younger of the two.

————

2*nd race*, 2.5 p.m.

Distance, 7 furlongs.
Time occupied in running, 1 minute 42$\frac{2}{5}$ seconds.
" Off," 2.7
Finishes 2.8.42$\frac{2}{5}$ p.m., under " 9."

Time Table.

6

3   9   (4.1)

2.7   2.11   2.15

The first three horses—

Fairy Field	.	$424 = \frac{1}{3}$	6 years old.
Black Mail	.	$133 = \frac{7}{8}$	3	,,
Lucinda .	.	$145 = 1$	4	,,

3rd race, 2.50 p.m.

Distance, 6 furlongs.
Time occupied in running, 1 minute $10\frac{4}{5}$ seconds.
" Off," 3.1.
Finishes 3.2.$10\frac{4}{5}$ p.m., under " 3."

Time Table.

5

5   7.2   8   3   9

2.51   2.55   2.59   3.3   3.7

The first three horses—

Cossack .	.	$102 = 3$	4 years old.
Master Willie	.	$747 = \frac{9}{1}$	6	,,
Indian Corn	.	$387 = \frac{9}{2}$	5	,,

This requires no explanation at all.

4th race, 3.30 p.m.

Distance, about $1\frac{1}{2}$ mile.
Time occupied in running, 2 minutes $41\frac{2}{5}$ seconds.
" Off," 3.46.
Finishes 3.48.$41\frac{2}{5}$ p.m., under " 7.2 " (?).

Time Table.

5
3   9   (4.1)   6   5   7.2
3.31 3.35 3.39 3.43 3.47 3.51

The first three horses—

Osboch .	.	86 = 5	4 years old.
Volodyovski	.	222 = 6	„
Santoi .	.	523 = 1 6	„

Curiously enough, the result shows that the race ought to have been finished under the 5 period. The case seems somewhat abnormal, but it is possible that a mistake of one minute may have been made in the record of the time. In any case the numbers of the placed horses are significant of the regularity of the law's operation.

———

5*th race*, 4.5 p.m.

Distance, 5 furlongs.
Time occupied, about a minute.
" Off," 4.21.
Finishes 4.22 p.m., under " 8."

Time Table.

(7.2)
(4.1)   6   5   (7.2)   8
4.7   4.11 4.15 4.19 4.23

The first three horses—

Regalia .	.	262 = 1	3 years old.
Egmont .	.	123 = 6	5 „
Wisconsin	.	248 = 5	3 „

Under the law of " interchange," 1 wins under the influence of 8.

———

6*th race*, 4.40 p.m.

Distance, 1¼ mile.
Time occupied, 2 minutes 8 seconds.
" Off," 4.55½.
Finishes 4.57.38 p.m., under " 9."

Time Table.

		8		
(7.2)				
5	(7.2)	8	3	9
4.43	4.47	4.51	4.55	4.59

The first three horses—

Baldoyle . . 86 = 5 6 years old.
Ypsilanti . . 632 = 2 4 „
Australian Star . 1414 = ¼ 6 „

Law of interchange being applicable, 5 wins in 9's time.

———

ASCOT—*Tuesday*, the 17*th June* 1902 (Sun rises, 3.44).

1*st race*, 1.30 p.m.
    Distance, about a mile.
    Time occupied in running, about 2 minutes.
    " Off," 1.46.
    Finishes about 1.48, under " 5."

Time Table.

6					
3	9	(4.1)	6	5	7.2
1.32	1.36	1.40	1.44	1.48	1.58

The first three horses—

Rose Blair . 455 = 5 3 years old.
Simoom . . 146 = 2 3 „
Compliment . 672 = 6 4 „

It is so clear that no remarks are necessary.

———

2*nd race*, 2 p.m.
    Distance, about 2 miles.
    Time occupied in running, 3 minutes 50 seconds.
    " Off," 2.16.
    Finishes 2.19.50, under 7.2 apparently, but really
        under " 8."

Time Table.

<pre>
5
9   (4.1)   6   5   (7.2)   8
2.4   2.8   2.12   2.16   2.20   2.24
</pre>

The first three horses—

Ice Maiden   .   166 = $\frac{4}{8}$ 3 years old.
Prince Florizel .   753 = $\frac{6}{3}$ 3   „
Rice   .   .   260 = 8   5   „

Ten seconds difference takes the race under " 8," where Ice Maiden correctly wins; the placed ones point to this fact plainly.

————

3rd race, 3 p.m.

Distance, 5 furlongs 136 yards.
Time occupied in running, 1 minute 18 seconds.
" Off," 3.8.
Finishes 3.9.18 p.m., under " 5."

Time Table.

<pre>
(7.2)
(4.1)   6   5
3.4   3.8   3.12
</pre>

The first three horses—

Rocksand   .   .   336 = $\frac{3}{5}$ 2 years old.
Baroness La Fleche .   780 = 6   „
Red Lily .   .   .   284 = 5   „

The result being plain, no explanation is necessary.

————

4th race, 3.30 p.m.

Distance, about 2 miles.
Time occupied in running, 3 minutes 41$\frac{3}{5}$ seconds.
" Off," 3.47.
Finishes about 3.51 p.m., under " 3."

Time Table.

(7.2)
(4.1)   6   5   7.2   8   3
3.32   3.36   3.40   3.44   3.48   3.52

The first three horses—

Scullion   .   .   170 = 8   4 years old.
Carbine   .   .   283 = 4   4   „
Rambling Katie   774 = $\frac{0}{8}$ 5   „

Scullion wins near its own group by the strength of its being a *compared number*, as has been explained before. Last year this race was won by Sinopi, 206 = 8, being the same number and the same sound to which the winner of the present year belongs.

———

5*th race*, 4 p.m.

Distance, 5 furlongs 136 yards.
Time occupied in running, 1 minute 18$\frac{4}{5}$ seconds.
" Off," 4.25.
Finishes 4.26.18$\frac{4}{5}$ p.m., under " 4.1."

Time Table.

        8
        6   5   (7.2)   8   3   9   (4.1)
        4.4   8   12   16   20   24   28

The first three horses—

Quintessence   .   660 = 3   2 years old.
Kroonstad   .   741 = 3   „
Padilla   .   .   116 = 8   „

A little over a minute will take the race under " 6," where Quintessence wins under the law of " interchange." This fact is pointed to by the two companions.

6th race, 4.30 p.m.

> Distance, 1 mile 5 furlongs.
> Time occupied in running, 3 minutes $1\frac{4}{5}$ second.
> " Off," 4.54.
> Finishes 4.57.$1\frac{4}{5}$ p.m., under " 6 " apparently, but
> really under " 4.1," about a minute earlier.

Time Table.

8							
6	5	7.2	8	3	9	4.1	6
4.32	4.36	4.40	4.44	4.48	4.52	4.56	5.0

The first three horses—

Ard Patrick	.	$907 = \frac{7}{8}$	3 years old.
Perseus	. .	$410 = 5$	,,
Cheers	. .	$273 = 3$	,,

Again the result shows that there was probably a minute's error in the record of the time at which the race actually finished. Ard Patrick, as has been shown in case of the " Derby Stakes," wins under the influence of the Sun's time, his number, 8, being interchangeable therewith.

———

The above examples are quite sufficient to demonstrate the operation of the law explained in this book, but the reader is earnestly invited to examine as many instances as would actually satisfy him that the present ones are not selected unfairly.

If he should ever land himself in embarrassment, and fail to see that the great laws we are dealing with are in effectual operation, he may rest assured that either there is some error in the times recorded or in his own calculations, or sometimes possibly that sunrise time may have been incorrectly given in the almanac he has used. We have taken our " sunrise " from the Raphael's Almanac for 1902.

For the purposes of ready reference in cases of test, Appendix " A " will be found a most useful compilation.

CHAPTER V

PLANETARY SOUNDS

THE preceding chapters have embodied an attempt to show, by the practical application of the Science of Number to the records of the Turf, that a deep and profoundly mysterious law associates Number with the events of human life in a way that no modern speculation has hitherto perceived. In order to establish the fact of this relationship on a secure foundation, we have been content, so far, to state the rules which can be used to prove the existence of such a law, without going into the ramifications of the subject which may help to explain how those rules have been reached. The interest of the demonstration given will be twofold. For some readers probably the bearing of the law on the possibilities of practical divination, in a field of activity so intimately mixed up with worldly affairs as the race-course, will be the principal charm of these teachings. For others, the light they throw upon previously unsuspected influences playing on the affairs of the world, will be far more important than any value they may have in connection with betting operations. Our present task, therefore, will be to discuss the significance of the laws observed, from the point of view of the metaphysical student seeking to understand, as far as the conditions of life on this physical plane will allow, the extent to which the apparent freedom of the human will is curtailed by invisible forces which lead it into the paths along which it actually travels. Probably, for that matter, the revelations of previously hidden knowledge which this book contains are little more than

5

a few fragments of the real Science of Number which ancient wisdom has always pointed to as embodying the principle of creation in some unexplained way ; but, as far as they go, their significance is very wonderful, and they tempt the thoughts of earnest students of Nature along many lines of philosophical speculation that have hitherto proved very barren.

Mention has been made already of the circumstances under which the author was tempted into the line of inquiry that led to his present conclusions. The use of the 4-minute periods was only suggested by a very prolonged observation of the facts to be analysed, but it is not to be thought of as an arbitrary selection. The period is one determined by Nature, discovered and not invented by the author. It is a natural period, derived, perhaps, from the fact that the Sun crosses over one degree of longitude every four minutes in the course of his (apparent) diurnal journey in the heavens.

Researches connected with the study of the Science of Number which the author has carried out at the public libraries of various foreign cities, as well as at the British Museum, have suggested many of the detailed methods of calculation he now employs, and the arrangement of the planets in the order of their influences through the hours of the day, and the short periods into which each hour is divided, is one that is found to have a very ancient sanction. It is also susceptible of something resembling an interpretation, for the grouping is harmonious in more ways than one.

	Negative.				Positive.	
☽	☿	♀	☉	♂	♃	♄
7.2	5	6	4.1	9	3	8
	20				20	

By natural addition each 20 is 2, and the two 2's make 4, the number of the Sun put in the middle. Furthermore, this arrangement harmonises with one which at the first glance appears quite incoherent with it, the order of the days of the week. It is obvious that the days of the week are named after the seven heavenly bodies in the above diagram. Some of the names in English have come down through Scandinavian mythology, but the identities of that with the mythology of Greece and of that with the planetary names is quite clear. Sunday, the Sun's day, is the first of the series, and the others in the accepted order are—Monday, the Moon's day; Tuesday, shown by the French name Mardi to be Mar's day; Wednesday, Woden's day, Woden being interchangeable with Mercury, as is further evidenced by the French name of the day, Mercredi; Thursday, Thor's day or Jupiter's; Friday, Venus's day (compare the French Vendredi); and Saturday, obviously Saturn's day. Now, looking at the above diagram, it will be seen that this order is derived from the central beginning of the series, Sunday, by a pendulous oscillation, the first swing going out to the farthest planet on the negative side, the Moon; then to the nearest on the positive side, Mars; then to the second of the negative series, Mercury; then to the second of the positive series, Jupiter; then to the third of the negative side, Venus; then, and lastly, to the third on the positive side, Saturn.

We do not say that this interpretation of the week is luminous with any particular significance, but it suggests an idea, as embodied in a series, which all students who have attempted hitherto to grapple with the subject have been wholly unable to associate with any coherent theory.

Why are some of the planets in the above diagram described as " positive " and the others as " negative " ? Volumes, representing ideas of a more or less hazy kind, have been devoted to the illumination or obscuration, as the case may be, of the fundamental principle of *duality*, running all through the operations of Nature. The male and female manifestations of humanity present the principle in its most glaring form, but philosophy detects the male and female principle in multitudes of natural manifestations that have nothing to do directly with animal organisation. Everything is coloured in some more or less obscure fashion with the one or other of the opposite, and therefore sympathetic, principles. Amongst the heavenly bodies the Sun stands pre-eminently the representative of the positive, active, creative principle. The Moon, in almost all languages treated as feminine, is pre-eminently representative of the receptive, passive, or negative principle. Both positive and negative, indeed, are equally creative, for creation is impossible unless both are concerned. But there are obviously positive and negative aspects of the creative force or principle. Now, as regards the days of the week, the pendulous oscillation goes alternately to the positive and negative planets. To some minds all such thoughts are void of meaning. They will not be able to discern the opposite principles, except in the very obvious case where it takes the shape of sex, but it is there in some mysterious disguise, and when we are trying, as in the present volume, to unravel some of the deeply seated and closely veiled laws of nature, we must take account of forces and principles that are very far from being obvious to the physical sight, and amongst these we must constantly be on the alert to recognise the characteristics indicated by the expressions under notice, positive and negative.

We may now pass on to a refinement connected with the planetary influences prevalent on the various days of the week, and during the minor periods, of which no notice has yet been taken. It is not necessarily brought into use in working out such calculations as those which determine (within certain limits) the winner of a race, but it can be superadded to those calculations as a check upon their accuracy, and, independently of this, will have a great deal of interest for readers who are studying the present volume for the sake of the light it casts on the working of Nature's laws,—not merely for what they may regard as its practical usefulness in worldly affairs.

Each planet is not merely associated with a number. It is associated with certain sounds susceptible of expression in writing. Thus it necessarily follows that such sounds have a greater value, or force, during the hours and periods governed by each planet in turn, than at other times. But now we are dealing with sounds themselves—not with their numerical values; and the practical effect of the state of things under notice is this. A name which *begins* with a certain sound is more intimately associated with some corresponding planetary influence than with any other. This influence would not override the indications to be obtained from the numerical value of a name, when these are plain and unmistakable, but may be very helpful in practice where the major indications are confused or difficult to understand. Some of the initial sounds with which we must now be concerned can be expressed by single letters; others by combinations of letters, as the following table will show :—

♄	♃	♂	☉	♀	☿	☽
S	F	A (as in "all" or "art")	M	A (in all cases not covered by the ♂ sound)	K	H (except in ☿ sound)
Sh	Ph	O (as in "orb")	T	E	C (hard)	D
C (soft)	Th	L		O (long)	P	
J	Y (when followed by a nasal sound, as in "young")	N		B	H (followed by broad "a")	
G	Cha (as in "charm")	Y (in all cases except the preceding)		R	O	
Z	Chi	Choo		V	Toe	
X (when sounded like Z).	Che (as in "cheese")	Che (as in "cherry")		W		
		Cho (as in "choke")		I (short)		
		U				
		I (long)				

Now, it is a matter of common observation among people who pay attention to racing records that a

majority of the races on any given day will often fall
to horses whose names begin with the same letter.   In
ordinary life this state of things, when noticed, is merely
regarded as an amusing coincidence, but anyone who
takes the pains to apply the table just given to the
records in question will discover that the extent to which
one or other of the sounds associated with some particular
planet is found to begin the names of most of the winning
horses on any given day, entirely outruns the " chance "
of coincidence.   Or the mysterious influence in question
will be found to operate transversely to the flow of time
during any day; that is to say, all the races run at
3 o'clock (for example) on the successive days of a long
race meeting will fall to horses with the same planetary
sound as the initial sound of their names.   The influence
which this state of things represents may, as we have
said, be overridden by stronger influences, or blended
with these so as to modify their effects, but no one can
rise from a patient survey of the facts without being
convinced that there is a great deal more in this law
than meets the eye at the first glance.   It is easy to
run through a few records to show what we mean more
clearly.

Let us take a *M'Call's Racing Chronicle* for last year,
and, to avoid all suspicion of picking illustrations to suit
our purpose, let us begin at the beginning.   The first
meeting recorded is that of Nottingham, in the November of the previous year.   The first race results as
follows—

Clansman   1
Kurvenal   2

Both names begin with the sounds belonging to Mercury.

The second race is under the Venus influence, the
winning names being—

Erich
Baldoch
Balantine.

The third race still shows the same influence—

Valdis	I
Rapine	2

The fourth—

Halutos	I
Hornpool	2

Both are names beginning with Moon sounds.

At the same place next day the first race is won by Lord Foppington. In other calculations we generally ignore the mere title; but in observing the planetary sound influences, it seems as though, more often than not, the title must be taken into account. There is no other horse in the race with Lord Foppington with a name belonging to the L or Mars series, but in the second race both first and second—

Miss Royston	I
Margaret	2

belong to the Sun series.

In the third race the record is as follows—

Ortygian	I
Martin	2
Odran	3

Here the second horse seems to give us the first case showing a failure in the operation of the mysterious law. The first and third are Venus names, while the second is a Sun name. But this is simply a case in which the numerical influence has overridden the other. As we have said already, the law of number is the primary force to be considered, and the initial sound law is found to harmonise with it so frequently in practice as to be worth earnest attention; but it is not in itself overwhelming.

The fourth race of the day still under notice is won by—

Alpheus 1
Blue Mint 2
Opae 3

all Venus names.

At Folkestone, on the next page of our *Chronicle*, we find the following races vindicating the mysterious law—

Corner 1
Campana 2
Chair of Kildare 3

In this and succeeding examples we leave the reader to observe for himself the identities of the sounds by reference to the table already given.

Tom Tit 1
Moonfish 2

(Wood) Pigeon 1
(Fool's) Paradise 2

Here it seems as though we have to ignore the first part of each name. The reader must remember that we are not yet declaring any rule by which it is always possible to determine how to manage double-barrelled names. The point now emphasised is that obviously a law is at work, though we may not yet be in command of all the collateral laws which now and then modify its operation. To resume—

George Fordham 1
Germanicus 2

For a change, let us take a few striking examples from the *Chronicle* relating to the first half of the present year, and including the later meetings of 1901. See the Maiden Erlegh meeting of November—

Marriage Lines 1
Tarolinta 2
Molester 3

See Kempton Park, November—

Overrated	.	.	.	.	I
Oban	.	.	.	.	2
Exccpcionale	.	.	.	.	3
Tonsure	.	.	.	.	I
Traveller	.	.	.	.	2

See Wye meeting, December—

Kurvenel	.	.	.	.	I
Peopleton	.	.	.	.	2

See Shirley Hunt—

Postman's Knock	.	.	.	I
Plumage	.	.	.	2

See Manchester, December—

Stirtloe	.	.	.	.	I
Glory Hole	.	.	.	.	2
Greek Lass	.	.	.	.	3

See Hurst Park, January—

Spinning Boy	.	.	.	I
Shackleford	.	.	.	2
Goldwasher	.	.	.	3

It is unnecessary to load these pages with further examples. The reader can go for himself, if he likes to take the trouble, over any racing records, and he will find our law operative—not in absolutely every case, but in so many that all theory of chance in the matter is utterly ruled out of court. As we have called it already, it is a very mysterious law, and will repay a great deal more painstaking study than the author of this volume has yet been able to devote to it, but its interest from the higher point of view of occult study is beyond question.

The principle which we thus perceive to be at work

is not sufficiently commanding to be trusted by itself as forecasting the result of a competition, but when it is seen to operate, concurrently with the significance of a numerical calculation, it may be recognised as giving a greatly enhanced value to that calculation. Suppose, that is to say, that in trying to work out the probability of two races from the names of the horses concerned, the result in the one case pointed to the success of horses whose names began with the planetary sound corresponding to the periods in which the issue of the race was expected; while, in the other case, there was no such correspondence, the student might feel quite sure that his calculations were right in the first case, and might reasonably apprehend that he had made some mistake (probably in counting the values of the names) in the other.

And besides the use which may be made of the planetary sound in checking numerical calculations, any-one who can remember the table just given has a very useful rough test applicable to the probabilities of any such competition as a race. When the blind preferences of the betting ring are seen to be at variance with the indications of planetary sounds, the likelihood that the "favourite" will be beaten is very clearly pronounced for the observer armed with the hitherto occult knowledge now disclosed. It does not follow that the planetary sound test would in such cases point out the winner. It might, if there happened to be no other name concerned except one which began with the right sound, but more probably there would be more than one so qualified. And it is not always easy to determine whether the planetary sound influence is most effective along a series of events on any one day, or along the corresponding hours of successive days. The charm and interest of the matter have to do with the glimpse it affords the occult student of a law in operation, which as yet he must be content to understand imperfectly, but which he will not be the less encouraged to study on that account.

Indeed, as we have remarked from time to time in the course of the preceding pages, the importance of appreciating the manner in which the life of human beings on this world is wrapped round with all manner of controlling forces entirely unsuspected by the races of the West, engrossed hitherto by their study of *physical* nature, is the explanation of the production of this book. The realisation of this great truth should exert a wholesome influence on the mind, as we will endeavour presently to show; but it is also liable to create an uncomfortable feeling if it is misunderstood. Some of us may be disagreeably startled at finding results they have been in the habit of setting down to their own action and choice, due apparently to some superior control, the nature of which has hitherto been entirely concealed from their observation. The race-horse owner, for example, has been in the habit of attributing his success, when he obtains success, to his own skill in breeding and training his animals, and has been fully assured that as far as the insignificant matter of naming them has been concerned he has absolutely followed the initiative of his own fancy. It is bewildering for him, if he sees the force of the demonstrations this volume contains, without seeing something more, to find that he and his horses are somehow the toys of an influence too subtle to have been previously dreamed of, and the discovery may induce him to regard all his own efforts in the direction of securing success as so much energy thrown away. That is not the true moral to be derived from the experience the reader has gone through, and the system we have set forth. Human action is an all-important factor in the development of the conditions which the science of number enables us to detect when they exist. We do not say that in the case of any given race, started at such a moment on such and such a day, the number of the horse's name is the *cause* of his victory. But, owing to the marvellous manner in which Nature contrives to blend a great many converging streams of influence, it will always happen that the horse qualified by his muscles and training to win *will have* a name number that chimes in with the planetary influence operative when the race is actually run. The principle we are now emphasising is perhaps even more plainly manifest in the case of competitive examinations. There the numerical value of the candidate's name is assuredly not the *cause* of his triumph over his rivals. That is due, as far as immediate relationships of cause and effect are concerned, to his brain capacity, his industry as a student, his state of health, nervous organisation, and so forth. Adopting the language of some Eastern schools of philosophy it is his *karma* that really accounts for his success, the sum total, that is to say, of his action in former lives and in this. But Nature has blended with his karma various characteristics in harmony therewith. *It will happen* that the struggle of his competition will culminate at a period when the planetary influences that correspond with the numerical value of his name are predominant.

No teaching could be more immoral than one which seemed to show human beings absolutely the toys of destiny in regard to all their doings, great and small. It would take away all motive for wholesome exertion, and lay an axe at the root of all our conceptions of right and wrong. Such teaching would be absurd in the sight of all who have even a glimmering perception of the principles that are guiding human evolution. But no less absurd would it be to imagine that the progress of the human race as a whole, or the fate of the world, is left to the uncertain influence of human action, illuminated by no higher knowledge than that which prevails even in the most civilised communities at the present day. Our humanity as a whole is still very young, and is held in leading strings not the less trustworthy or unbreakable for being quite unperceived in most cases by those whom they guide or hold in check. There is just enough elasticity in these leading strings, or just enough "slack" in the lines, if we adopt another metaphor, to leave each of us free to manifest and develop the tendencies of his nature. But to mistake this limited freedom for independence of all superior control is to misunderstand the whole scheme of Providence. It is exactly by making this mistake that so many highly cultivated European thinkers have drifted into the belief that the world of matter they see, with its mysterious potentialities of life, is a complete thing in itself, connected with no other scheme of existence above or below, so that each life when finally snuffed out is done with for evermore. No doubt the tide of materialistic thought which swept with such force through the latter half of the last century has now been checked and turned back by the influences of teaching and experience in various forms, putting a scientific face upon spiritual conceptions that had lost the hold they had on the human mind during the "ages of faith." But all the more has it become desirable to bring the invisible world into such relationship with everyday life, that the modern world shall be unable to pretend that it disbelieves in the influence of invisible intelligence on the practical concerns of man. The mediæval worshipper at the shrine of a saint needed no proof to make him feel sure of a power above him which it was his duty to reverence. And his mind would not have been receptive of any ideas concerning that power that were advanced beyond the crudities of the monkish tradition. But the modern sceptic is cast in a very different mould. The theology of the Middle Ages has amused or bored him, as the case may have been. The intellectual triumphs of his time have been concentrated on the phenomena of the material universe. He is simply unacquainted with any facts within the range of experience that seem to link these phenomena with extra-terrestrial planes of existence. For some persons, indeed, who have been in that attitude of mind, the various developments of super-physical research in progress for the last dozen years or so have been extremely significant,—in many cases quite conclusive. But there is still a great deal to be accomplished before the public beliefs of the time can be guided, to any comprehensive extent, into the channels of thought which lead to a loftier wisdom. The enlightened masses—those who have assimilated the intellectual ideas of their generation, but have failed to find in any religious teaching the assurance of super-terrestrial influences in touch with our own world and our own daily lives—are in want of some experience of a sort that they can readily test and handle, that will settle the question about some unseen government of the world in an indisputable fashion. Over-refined argument would hardly help them ; other people's experience is of no account. This volume has been written to offer them the means of getting experience for themselves in connection with one of the pursuits most intimately associated with the lives of the greatest number. The puritans of occultism—for there are fanatics in every department of spiritual growth—will perhaps be shocked at the application of sacred mysteries to a pursuit so little sanctified in its general associations as horse-racing. But in a loftier than the usual cynical sense we may quote the saying, *Qui veut la fin, veut les moyens.* If we want to show our worldly-minded, pleasure-seeking friends that there is a complicated Providence ruling even their pursuits, let us do it in the only way that is effective,— by offering them a proof of a kind that can be appreciated and understood.

APPENDIX A

ALPHABETICAL LIST OF THE NAMES OF RACE-HORSES IN GREAT BRITAIN, SHOWING THE FULL VALUE, REMAINDER, AND AGE.

Note.—F. stands for Filly, C. for Colt, G. for Gelding, S. for Stallion, M. for Mare, and H. for Horse.

Name.	Sex.	Value.	Remainder.	Age (figures indicate years and "A" aged).	Remarks.
Abbess . . .	F.	73	1	4	
Abbey Bell . .	M.	65	2	A.	Bell, 42=6.
Abbeystead . .	C.	497	2	3	
Abbeywood . .	H.	33	6	6	
Abbot . . .	G.	413	8	5	
Abbots Anne .	F.	532	1	3	Anne, 61=7.
Abercorn (By) .	F.	475	7	2	
Abernethy .	F.	678	3	2	
Aborigine (By) .	F.	262	1	3	
Absent Friend .	F.	867	3	2	Friend, 344=2.
Absolution . .	F.	475	7	2	
Accountant . .	G.	927	9	A.	
Accuracy . .	F.	307	1	3	
Achaicus . .	C.	111	3	3	
Achates . .	C.	491	5	4	
Achray . .	M.	241	7	A.	
Acquisition . .	F.	388	1	2	
Activity . .	M.	921	3	6	
Actuary . .	C.	251	8	4	
Ada Cerito . .	F.	301	4	3	Ada, 16=7.
Adam II.. .	H.	45	9	A.	
Adamhill .	G.	80	8	A.	
Adansi . .	G.	126	9	4	
Adare II.. .	G.	215	8	A.	
Adavoyle . .	C.	138	3	4	

Name.	Sex.	Value.	Remainder.	Age (figures indicate years and "A" aged).	Remarks.
Adelina . . .	M.	96	6	5	
Aderno . . .	G.	271	1	3	
Adieu (By) . .	F.	21	3	2	
Admiral Dewey .	H.	325	1	5	Dewey, 21 = 3.
Aemilia . . .	F.	92	2	2	
Aeneas . . .	H.	132	6	5	
Aerie . . .	M.	221	5	A.	
Aerolite II. . .	H.	647	8	A.	
Aesculapius . .	G.	272	2	5	
Affluent . . .	F.	571	4	3	
Afonwen . . .	M.	203	5	6	
Agathos II. . .	G.	493	7	A.	
Agrimony . .	G.	331	7	3	
Aggressor . .	C.	501	6	3	
Agnes Forager .	F.	636	6	4	Agnes, 141 = 6.
Aigrette . . .	F.	641	2	2	
Airion . . .	C.	271	1	4	
Alan . . .	G.	91	1	4	
Alarm Bell . .	G.	314	8	6	Bell, 42 = 6.
Albemarle . .	C.	314	8	4	
Albert Edward .	G.	868	4	A.	633 = 3.
Albina . . .	F.	94	4	4	
Alboin . . .	C.	85	4	4	
Alcaline . . .	F.	141	6	4	
Alcibiades . .	H.	178	7	6	
Alcmena . . .	F.	152	8	3	
Alcove . . .	G.	137	2	5	
Aldbourne Chimes .	G.	397	1	A.	Chimes, 103 = 4.
Alderman . .	C.	326	2	2	
Alderney . . .	G.	296	8	5	
Alderwood . .	G.	246	3	A.	
Aleda . . .	F.	46	1	4	
Alencon . . .	C.	161	8	3	
Alethe . . .	F.	456	6	2	
Alexandria . .	M.	334	1	A.	
Alfar . . .	G.	312	6	A.	
Alfio . . .	G.	127	1	A.	
Ali . . .	C.	110	2	4	
Ali . . .	G.	110	2	A.	
Ali II. . . .	G.	110	2	3	
Alice Holt . .	F.	542	2	3	
Alien . . .	G.	101	2	4	
Alistra . . .	F.	692	8	3	
Aliwal . . .	G.	147	3	5	

Name.	Sex.	Value.	Remainder.	Age (figures indicate years and "A" aged).	Remarks.
Allesby . . .	H.	113	5	6	
All Hot . . .	C.	439	7	2	Hot, 407 = 2.
Allithorne . .	M.	688	4	5	
All Moonshine (By).	F.	478	1	2	Moonshine, 446 = 5.
All my Eye . .	F.	93	3	2	Eye, 11 = 2.
All Sunshine . .	M.	492	6	5	Sunshine, 460 = 1.
All's Well II. . .	G.	138	3	A.	
Allurement . .	F.	747	9	4	
Almanac . . .	M.	143	8	5	
Alone in London .	G.	272	2	A.	Alone, 87 = 6.
Alpheus . . .	G.	181	1	A.	
Alston's Pride . .	G.	885	3	4	
Alswitha . . .	M.	503	8	A.	
Altair . . .	H.	641	2	A.	
Altesse Royale . .	M.	743	5	A.	Altesse, 501 = 6.
Althotas (By) . .	G.	903	3	2	
Altnabreac . .	C.	714	3	4	
Alvaston Belle . .	F.	664	7	2	Belle, 42 = 6.
Alveston . . .	M.	631	1	6	
Alyssum . . .	M.	131	5	A.	
Amazonia . . .	F.	116	8	3	
Amazonius . .	C.	175	4	4	
Ambiguity . .	C.	473	5	3	
Amberite . . .	G.	653	5	A.	
Ambush II. . .	G.	343	1	A.	
American . . .	C.	331	7	4	
American Girl . .	F.	581	5	4	Girl, 250 = 7.
Amerique . . .	F.	281	2	4	
Amnesty . . .	G.	571	4	5	
Amer Picon . .	C.	403	7	4	Amer, 251 = 8.
Amoret II. . .	F.	657	9	3	
Amphictyon . .	G.	601	7	4	
Amphiery . .	F.	341	8	4	
Amphion (By) . .	C.	181	1	2	
—— . . .	F.	181	1	2	
—— . . .	G.	181	1	2	
—— . . .	C.	181	1	2	
Amphlett . . .	M.	571	4	5	
Amplify . . .	F.	251	8	2	
Amport . . .	F.	727	7	4	
Amulet . . .	M.	497	2	A.	
Amurath . . .	H.	647	8	6	
Amusement . .	M.	564	6	5	
Amy	F.	61	7	4	

Name.	Sex.	Value.	Remainder.	Age (figures indicate years and "A" aged).	Remarks.
Anahilt . . .	C.	487	1	3	
Ananias . . .	G.	172	1	A.	
A. N. B. . . .	H.	84	3	6	
Anchovy . . .	H.	150	6	A.	
Andrea Ferrara .	C.	758	2	4	Andrea, 266=5.
Angel . . .	F.	94	4	4	
Angel Court . .	C.	720	9	3	Angel, 4.
Angelet . . .	C.	504	9	2	
Angus . . .	G.	141	6	A.	
Ankerdine . .	G.	345	3	A.	
Anna Craig . .	M.	312	6	5	Anna, 62=8.
Annagor (By) . .	G.	282	3	4	
Annaross . . .	M.	328	4	6	
Annie Hermit . .	F.	716	5	4	Annie, 71=8.
Annual . . .	M.	97	7	5	
Annuity . . .	F.	477	9	2	
Ansley . . .	G.	161	8	3	
Antidote . . .	G.	881	8	A.	
Antrim Lass . .	F.	792	9	4	Lass, 91=1.
Anxious . . .	C.	441	9	4	
Aperfield . . .	H.	405	9	5	
Aperse . . .	F.	351	9	2	
—— . . .	F.	351	9	2	
Apis . . .	M.	141	6	A.	
Apollo III. . .	G.	41	5	A.	
Applause . . .	F.	42	6	4	
Applause II. . .	F.	42	6	4	
Appleblossom . .	F.	177	6	4	Blossom, 134=8.
Apple Tree . .	C.	653	5	2	Tree, 610=7.
April Fool II. . .	G.	437	5	6	Fool, 116=8.
Aqua Marine . .	C.	328	4	3	Aqua, 28=1.
Arabi . . .	C.	282	3	2	
Arab Maid . .	M.	326	2	5	Maid, 54=9.
Arachne . . .	M.	282	3	5	
Aralia . . .	F.	243	9	4	
Aratus . . .	C.	662	5	3	
Arbigland . .	F.	308	2	4	
Archon . . .	C.	272	2	2	
Archduke II. . .	G.	245	2	A.	
Archer . . .	G.	405	9	6	
Ardandra . . .	F.	461	2	5	
Ardeer . . .	H.	415	1	5	
Ardenheim . .	G.	310	4	4	
Ardeshir . . .	H.	715	4	A.	

Name.	Sex.	Value.	Remainder.	Age (figures indicate years and "A" aged).	Remarks.
Ardgreagh . .	M.	435	3	A.	
Ard Patrick . .	C.	907	7	3	Patrick, 701＝8.
Ardragh . . .	G.	426	3	3	
Area Belle . .	F.	264	3	2	Belle, 42＝6.
Argovian . .	F.	367	7	4	
Argument . .	G.	647	8	6	
Ariette . . .	F.	611	8	3	
Ariostro . .	C.	276	6	2	
Aristo . . .	C.	276	6	4	
Aristocrat . .	H.	897	6	5	
Arizona . . .	C.	265	4	3	
Arizona II. .	C.	265	4	3	
Arklows Pride .	G.	549	9	A.	
Arksey . . .	F.	292	4	3	
Arley . . .	G.	241	7	3	
Armadillo . .	G.	282	3	4	
Armeath . .	C.	656	8	4	
Armenia . .	M.	312	6	A.	
Armenian . .	H.	361	1	A.	
Armine . .	M.	301	4	6	
Armoy . . .	M.	253	1	5	
Arnold . . .	G.	286	7	A.	
Arriago . .	M.	238	4	5	
Arrogance . .	C.	347	5	2	
Arrowflight . .	F.	727	7	2	
Artiste . .	M.	1062	9	5	
Ascetic's Pride .	G.	835	7	A.	Pride, 284＝5.
Ascetic's Silver .	G.	921	3	5	Silver, 370＝1.
Asgarby . .	M.	294	6	A.	
Ashanti Gold .	C.	822	3	2	Gold, 60＝6.
Ashbower . .	F.	521	8	3	
Ashburn . .	G.	563	5	A.	
Ashling . .	G.	411	6	A.	
Ashtwig II. .	G.	346	4	A.	
Aslingdon . .	H.	225	9	6	
Asperity . .	F.	761	5	2	
Assault . . .	G.	493	7	4	
Assiout . .	C.	477	9	3	
Astronome II. .	C.	763	7	3	
Astronomer II. .	G.	953	8	A.	
Atalanta . .	F.	883	1	3	
Atalanta . .	F.	883	1	4	
Athel Brook . .	G.	674	8	4	Athel, 446＝5.
Athelfrith . .	G.	1131	6	A.	

Name.	Sex.	Value.	Remainder.	Age (figures indicate years and "A" aged).	Remarks.
Athelgold . . .	F.	506	2	4	
Atheling's Pride .	G.	807	6	A.	Pride, 284 = 5.
Atheling's Son .	C.	633	3	4	Son, 110 = 2.
Athel Roy . .	G.	658	1	A.	Athel, 446 = 5.
Athelwulf . . .	G.	562	4	A.	
Atholia . . .	F.	453	3	3	
Atrocious . . .	G.	967	4	3	
Attractor . . .	G.	972	9	A.	
Atty's Pride . .	G.	712	1	6	Pride, 284 = 5.
Aucasia . . .	M.	99	9	A.	
Auchnafree . .	C.	943	7	3	
Audiern . . .	C.	266	5	4	
Aughamore . .	G.	269	8	6	
Aughrim (By) . .	F.	262	1	2	
Auld Religion . .	M.	334	1	A.	Religion, 293 = 7.
Aunt Anne . .	M.	513	9	5	Anne, 61 = 7.
Aunt Ashe . .	M.	763	7	A.	Ash, 311 = 5.
Aunt Mary Anne .	F.	764	8	4	Mary, 251 = 8.
Aunt May . .	M.	502	7	6	May, 50 = 5.
Auratum . . .	C.	648	9	3	
Aureolus . . .	H.	313	7	6	
Ausonia . . .	M.	129	3	6	
Austerlitz . . .	G.	1099	1	4	
Australian Homer .	G.	1004	5	6	Homer, 251 = 8.
Australian Star .	H.	1414	1	6	Star, 661 = 4.
Autocrat (By) . .	C.	638	8	3	
Aventure . . .	F.	344	2	3	
Averse . . .	G.	341	8	A.	
Aversion . . .	F.	631	1	2	
Avincourt . . .	C.	757	1	3	
Avington (By) . .	F.	601	7	3	
—— . . .	C.	601	7	2	
Avoca Vale . .	F.	228	3	3	Avoca, 108 = 9.
Avonbeg . . .	M.	163	1	6	
Awakening (The) .	C.	171	9	2	Awakening, 157 = 4.
Away West . .	C.	493	7	3	Away, 17 = 8.
Ayacanora . .	M.	290	2	A.	
Ayala . . .	C.	43	7	4	
Aylsham . . .	G.	381.	3	4	
Aymon . . .	C.	101	2	2	
Ayrshire (By) . .	C.	711	9	3	
—— . . .	G.	711	9	2	
—— . . .	F.	711	9	2	
—— . . .	C.	711	9	2	

Name.	Sex.	Value.	Remainder.	Age (figures indicate years and " A " aged).	Remarks.
Ayrshire Beauty .	F.	1139	5	2	Beauty, 428 = 5.
Azaliel . . .	H.	79	7	6	
Azro	H.	214	7	5	
Babel . . .	M.	35	8	6	
Babworth . . .	G.	616	4	4	
Baby King . .	G.	105	6	6	King, 90 = 9.
Baby Lamb . .	F.	86	5	2	Lamb, 71 = 8.
Bacchante . .	F.	483	6	3	
Bacchus . .	G.	83	2	A.	
Bachelor's Button .	C.	748	1	3	Button, 452 = 2.
Bachelor's Fancy .	F.	497	2	2	Fancy, 201 = 3.
Bachelor's Pride .	G.	580	4	4	Pride, 284 = 5
Baden . . .	H.	57	3	6	
Bad News . .	F.	80	8	4	News, 73 = 1.
Bad Times . .	M.	507	3	A.	Times, 500 = 5.
Baggara . . .	G.	224	8	4	
Bagman . . .	G.	113	5	3	
Bahr Yousouf . .	C.	360	9	3	Yousouf, 156 = 3.
Bakersfield . .	G.	407	2	3	
Bala	G.	34	7	A.	
Baladine . . .	M.	98	8	6	
Balausta . . .	M.	500	5	6	
Baldoyle . . .	G.	68	5	6	
Baldur . . .	H.	237	3	6	
Ballantrae . .	F.	695	2	3	
Ballasalla . .	G.	126	9	3	
Baller Trough . .		839	2	6	Baller, 233 = 8.
Ballet Girl II. . .		293	5	5	Girl, 250 = 7.
Ballinagarde . .	G.	309	3	A.	
Ballinclair . .	G.	343	1	4	
Ballinterry . .	G.	703	1	6	
Ballista . .	G.	494	8	6	
Ballnagarde . .	C.	309	3	4	
Ballybrophy II. .	G.	337	4	A.	
Ballyconra . .	F.	320	5	4	
Ballyhaunis . .	M.	164	2	A.	
Ballylopen . .	C.	209	2	3	
Ballymoney . .	G.	149	5	A.	
Ballymore . .		289	1	6	
Ballyrag II. . .	G.	264	3	A.	
Ballyrush . .	C.	543	3	3	
Ballytrent . .	G.	1103	5	6	
Ballyvillane . .	G.	204	6	6	

Name.	Sex.	Value.	Remainder.	Age (figures indicate years and " A " aged).	Remarks.
Ballywalter . .	G.	650	2	4	
Balm of Gilead .	M.	189	9	A.	Balm, 44 = 8.
Balrath . . .	G.	639	9	4	
Balsarroch . .	C.	315	9	4	
Balsarroch II. .	G.	315	9	5	
Baltimore. . .	G.	679	4	4	
Banker . . .	G.	273	3	A.	
Band of Hope .	G.	229	4	6	Band, 57 = 3.
Bandparts . .	M.	798	6	A.	
Banjo II. . .	G.	62	8	A.	
Banjo III. . .	G.	62	8	5	
Bansha . . .	C.	354	3	4	
Barbara . . .	M.	406	1	A.	
Barbara Frietchie .	F.	709	7	4	Barbara, 406 = 1.
Barba Rossa .	G.	669	3	A.	Barba, 206 = 8.
Barberstown .	F.	921	3	4	
Barbuda . .	C.	210	3	4	
Barde . . .	C.	207	9	3	
Bards Gallop .	G.	398	2	3	Gallop, 131 = 5.
Barford . .	H.	487	1	A.	
Barham . . .	G.	248	5	3	
Barlock . .	F.	255	3	4	
Barnes Common .	C.	425	2	5	Common, 112 = 4.
Barona . . .	G.	260	8	A.	
Baroness La Fleche.	F.	780	6	2	Fleche, 420 = 6.
Baron Kendal .	H.	367	7	A.	Kendal, 114 = 6.
Baron's Folly .	G.	382	4	5	Folly, 122 = 5.
Barrackpore .	H.	503	8	5	
Barrackroom Pet	F.	959	5	3	Pet, 490 = 4.
Barrhill . .	G.	238	4	4	
Barrow II. .	G.	209	2	4	
Barry Sullivan .	G.	434	2	A.	Barry, 213 = 6.
Barsac . .	H.	313	7	A.	
Bar Scotch .	G.	288	9	5	Bar, 203 = 5.
Bar Sinister .	G.	973	1	2	Bar, 203 = 5.
Bashful . .	F.	413	8	2	
Baslow . .	G.	99	9	A.	
Basra . . .	F.	264	3	2	
Basse Terre .	F.	683	8	3	Basse, 63 = 9.
Bastard . .	G.	667	1	A.	
Bat . . .	C.	403	7	3	
Baton Rouge .	C.	959	5	3	Baton, 453 = 3.
Battalus . .	G.	102	3	A.	
Battels . .	F.	493	7	3	

Name.	Sex.	Value.	Remainder.	Age (figures indicate years and "A" aged).	Remarks.
Battledore . .	C.	643	4	2	
Battlement . .	C.	933	6	4	
Battleshiel . .	G.	773	8	4	
Battle Song . .	C.	575	7	3	Song, 132=6.
Bayfield . . .	G.	136	1	4	
Bayleaf . . .	M.	132	6	6	
Bay Ronald (By) .	F.	298	1	2	Ronald, 286=7.
—— . . .	G.	298	1	2	Ronald, 286=7.
—— . . .	G.	298	1	2	Ronald, 286=7.
. . .	F.	298	1	2	Ronald, 286=7.
Beaconsfield . .	G.	266	5	5	
Beatitude . .	M.	833	5	5	
Beatrice Maud .	M.	719	8	5	Beatrice, 673=7.
Beatrice R. . .	M.	875	2	6	Beatrice, 673=7.
Beau II. . .	C.	8	8	4	
Beau II. . .	G.	8	8	6	
Beauclerc (By) .	C.	268	7	2	
Beauty Bright .	F.	1030	4	3	Bright, 602=8.
Beauty of Kent .	M.	989	8	5	Beauty, 428=5.
Bebe . . .	C.	14	5	4	
Becky . . .	M.	42	6	6	
Becky Sharp .	M.	623	2	5	Becky, 42=6.
Bedhampton . .	G.	512	8	5	
Bedhu Husth .	G.	497	2	4	Bedhu, 27=9.
Bed of Heather .	F.	316	1	3	Bed, 16=7.
Bee Catcher .	G.	236	2	A.	Catcher, 224=8.
Beeswing III. .	M.	148	4	6	
Beetle . . .	H.	472	4	6	
Beggarman . .	G.	322	7	4	
Behind the Bush .	G.	387	9	A.	
Belamphion . .	G.	223	7	A.	
Belfast Boy .	C.	597	3	4	Boy, 14=5.
Belgrave (By) .	G.	352	1	A.	
Bel Item . .	M.	493	7	A.	Bel, 42=6.
Bella Angela .	M.	138	3	A.	Bella, 43=7.
Bella Gallina .	F.	145	1	4	Bella, 43=7.
Bellaria . .	F.	254	2	2	
Bellarmina . .	F.	334	1	4	
Bellarosa . .	M.	257	5	6	
Belle . . .	M.	42	6	5	
Belle Fille .	F.	162	9	4	Fille, 120=3.
Belle Magnifique .	M.	263	2	5	Belle, 42=6.
Belle of Braintree .	M.	975	3	5	Belle, 42=6.
Belle Promesse .	F.	438	6	4	Belle, 42=6.

Name.	Sex.	Value.	Remainder.	Age (figures indicate years and "A" aged).	Remarks.
Belle Poule . .	F.	158	5	2	Belle, 42=6.
Bellivor Tor . .	C.	534	3	3	Bellivor, 324=9.
Bell Sound . .	M.	162	9	A.	Bell, 42=6.
Belmeath . . .	H.	497	2	5	
Belmont . . .	G.	534	3	A.	
Beltenebrosa . .	F.	771	6	4	
Beluga . . .	F.	69	6	2	
Belvoir . . .	C.	292	4	3	
Ben Attow . .	C.	88	7	4	Attow, 26=8.
Ben Bolt II. . .	G.	500	5	A.	Bolt, 438=6.
Bend Or (By) . .	F.	268	7	2	
Benedictus . .	H.	556	7	A.	
Benella . . .	F.	103	4	2	
Benhead . . .	G.	81	9	5	
Beni Hassam . .	C.	408	3	3	Hassam (Arabic, Hasham), 346=4.
Benita . . .	M.	72	9	A.	
Benjamina . .	F.	156	3	4	
Benlight . . .	C.	492	6	2	
Bennitthorpe (By) .	F.	749	2	2	
Benoni . . .	M.	128	2	5	
Bentworth (By) .	F.	1073	2	2	
—— . . .	G.	1073	2	2	
Benvenir . . .	M.	402	6	6	
Bereaved . . .	C.	306	9	2	
Berehaven . .	H.	348	6	5	
Berengarins . .	H.	593	8	A.	
Beretta . . .	M.	232	7	A.	
Berne . . .	F.	252	9	3	
Berners . . .	G.	512	8	A.	
Bernicia . . .	M.	323	8	5	
Bessell . . .	M.	112	4	5	
Best-away . .	F.	489	3	3	
Bestbelle . . .	F.	514	1	2	Belle, 42=6.
Bestiole . . .	M.	518	5	5	
Best Man (By) .	F.	563	5	2	Man, 91=1.
—— . . .	F.	563	5	3	Man, 91=1.
Best of Three . .	M.	1168	7	A.	Best, 472=4.
Betsy . . .	F.	482	5	2	
Betsy . . .	M.	482	5	A.	
Betsy's Boy . .	G.	503	8	6	Boy, 14=5.
Betty Agnes . .	F.	563	5	3	Betty, 422=8.
Betty Snow . .	F.	538	7	2	Betty, 422=8.
Beverley Buck . .	G.	354	3	5	Buck, 22=4.

Name.	Sex.	Value.	Remainder.	Age (figures indicate years and "A" aged).	Remarks.
Bevil . . .	G.	122	5	A.	
Beware (By) . .	F.	228	3	2	
Bibury . . .	C.	244	1	2	
Biddie Hackler .	M.	272	2	A.	
Biddo . . .	G.	12	3	6	
Biddy . . .	F.	16	7	2	
Bide a Wee . .	M.	33	6	A.	
Big Busbie . .	G.	96	6	A.	Busby, 74 = 2.
Big Wheel . .	H.	68	5	5	Wheel, 46 = 1.
Bilbaude . .	M.	40	4	5	
Bill Garrett . .	G.	663	6	5	Bill, 32 = 5.
Bill Murphy . .	H.	362	2	A.	Bill, 32 = 5.
Billy Diver . .	G.	326	2	A.	Billy, 42 = 6.
Billy George . .	G.	250	7	6	Billy, 42 = 6.
Bindon . . .	H.	106	7	A.	
Binocle . . .	C.	104	5	3	
Binsonia . . .	F.	179	8	4	
Biology . . .	G.	63	9	5	
Bird	G.	206	8	A.	
Bird Call . .	M.	258	6	6	Call, 52 = 7.
Birdforth . .	F.	891	9	4	
Bird of Freedom (By)	C.	621	9	2	Bird, 206 = 8.
Bird of Paradise .	H.	580	4	5	Bird, 206 = 8.
Bird of Passage III.	M.	431	8	5	Bird, 206 = 8.
Bird on the Wing .	G.	348	6	A.	Bird, 206 = 8.
Bird's Eye . .	M.	224	8	A.	Eye, 11 = 2.
Birdsgrove . .	C.	572	5	2	
Birkacre . . .	G.	443	2	6	
Bisbille . . .	F.	104	5	3	
Bishop . . .	C.	382	4	4	
Bishop of Hereford .	C.	972	9	2	Bishop, 382 = 4.
Bistonian . . .	C.	187	7	4	
Bitter Almond .	C.	698	5	2	Almond, 96 = 6
Black and Star .	C.	769	4	2	Star, 661 = 4.
Blackbird . .	C.	259	7	3	
Black Bread . .	G.	269	8	6	Bread, 216 = 9.
Black Bush . .	G.	355	4	4	Bush, 302 = 5.
Black Cat . .	M.	474	6	6	Cat, 421 = 7.
Black Draft . .	G.	738	9	A.	Draft, 685 = 1.
Black Draught II. .	G.	738	9	A.	Draft, 685 = 1.
Black Fancy . .	F.	254	2	3	Fancy, 201 = 3.
Black Friar . .	C.	534	3	2	Friar, 481 = 4.
Blackguard . .	G.	278	8	A.	
Black Hugh . .	G.	74	2	4	Hugh, 21 = 3.

Name.	Sex.	Value.	Remainder.	Age (figures indicate years and "A" aged).	Remarks.
Black Hamburgh .	G.	321	6	6	Hamburgh, 268=7.
Black Jack . .	G.	77	5	A.	Jack, 24=6.
Black Lion . .	H.	143	8	A.	Lion, 90=9.
Black Mail . .	C.	133	7	3	Mail, 80=8.
Black Mark . .	C.	314	8	2	Mark, 261=9.
Black Meg . .	F.	123	6	2	Meg, 70=7.
Black Pearl .	F.	363	3	3	Pearl, 310=4.
Black Prince II. .	G.	443	2	A.	Prince, 390=3.
Black Sand .	H.	168	6	5	Sand, 115=7.
Blackthorn II. .	G.	710	8	5	
Black Watch II. .	G.	63	9	A.	Watch, 10=1.
Blagueur . .	G.	263	2	4	
Blair Anchor .	F.	523	1	2	Anchor, 281=2.
Blairgowrie .	C.	478	1	4	
Blakemere .	C.	312	6	3	
Blandford .	G.	371	2	A.	
Blantyre . .	G.	693	9	5	
Blarney II. .	G.	293	5	A	
Blazes . .	G.	47	2	5	
Blencathra .	F.	719	8	2	
Blend . .	H.	96	6	A.	
Blenheim . .	G.	147	3	A.	
Blesk . .	G.	122	5	5	
Blisworth .	G.	703	1	4	
Blitz (By) . .	M.	439	7	6	
Blondin II. .	G.	138	3	A.	
Bloomer . .	G.	278	8	A.	
Blossom . .	F.	136	1	4	
Blowing Stone .	G.	624	3	2	Stone, 516=3.
Blue Bridge .	G.	243	9	A.	Bridge, 205=7.
Bluecoat Boy .	C.	478	1	3	Boy, 14=5.
Blue Craigs .	C.	348	6	4	Craigs, 310=4.
Blue Diamond .	H.	136	1	5	Diamond, 98=8.
Blue Glass .	G.	149	5	5	Glass, 111=3.
Blue Gown II. .	F.	114	6	4	Gown, 76=4.
Bluegrass . .	F.	319	4	4	Grass, 281=2.
Blue Green (By) .	G.	318	3	4	Green, 280=1.
—— . .	F.	318	3	2	Green, 280=1.
—— . .	F.	318	3	3	Green, 280=1.
—— . .	C.	318	3	3	Green, 280=1.
—— . .	C.	318	3	3	Green, 280=1.
—— . .	C.	318	3	2	Green, 280=1.
—— . .	C.	318	3	2	Green, 280=1.
Blue Hare . .	C.	253	1	2	Hare, 215=8.

Name.	Sex.	Value.	Remainder.	Age (figures indicate years and "A" aged).	Remarks.
Blue Lamp . .	M.	189	9	A.	Lamp, 151 = 7.
Blue Look Out . .	F.	495	9	2	Look Out, 457 = 7.
Blue Melton . .	H.	568	1	5	Melton, 530 = 8.
Blue Mint . .	G.	528	6	A.	Mint, 490 = 4.
Blue Peter . .	C.	728	8	3	Peter, 690 = 6.
Blue Ruin . .	G.	294	6	5	Ruin, 256 = 4.
Blue Sleeves . .	C.	278	8	3	Sleeves, 240 = 6.
Blue Streak . .	C.	728	8	3	Streak, 690 = 6.
Blue Tyne . .	F.	498	3	3	Tyne, 460 = 1.
Blue Veil . . .	M.	128	2	A.	Veil, 90 = 9.
Blunt . . .	M.	482	5	A.	
Blythswood . .	C.	507	3	3	
Boa . . .	G.	9	9	A.	
Boathead . .	G.	427	4	A.	
Bobaire . .	F.	216	9	2	
Bobbie Burns . .	H.	328	4	A.	Bobbie, 16 = 7.
Bobette . .	M.	416	2	5	
Bob Major . .	G.	250	7	5	Bob, 6 = 6.
Bob Rynd . .	G.	260	8	6	Bob, 6 = 6.
Bobs . .		66	3	A.	
Bobs III.. . .	H.	66	3	5	
Bobsic . .	G.	76	4	A.	
Bob the Devil .	C.	144	9	4	Bob, 6 = 6.
Boddikins . .	F.	138	3	2	
Bodkin . . .	M.	78	6	A.	
Bolts and Bars .	F.	816	6	4	Bolts, 498 = 3.
Bonanza . .	F.	117	9	2	
Bonarcado . .	C.	290	2	4	
Bondi . . .	C.	68	5	2	
Bonfire . .	G.	344	2	A.	
Bonne et Belle .	F.	507	3	3	Bonne, 54 = 9.
Bonnet Rouge (By) .	F.	570	3	2	Rouge, 506 = 2.
Bonnie Chieftain .	H.	607	4	6	Chieftain, 543 = 3.
Bonnie Dundee .	G.	132	6	A.	Dundee, 68 = 5.
Bonnie Lassie .	M.	165	3	6	Lassie, 101 = 2.
Bonnie Pet .	F.	554	5	2	Pet, 490 = 4.
Bonnie Scotland .	C.	629	8	3	Scotland, 565 = 7.
Bonnie Skelton .	M.	634	4	5	Skelton, 570 = 3.
Bonny Common .	G.	176	5	A.	Common, 112 = 4.
Bonny Flora . .	F.	381	3	3	Flora, 317 = 2.
Bonny Yorkshire Lad	C.	831	3	4	Lad, 35 = 8.
Boomer . . .	H.	248	5	6	
Boothman . .	G.	503	8	4	
Boarder Boaster .	G.	1076	5	6	Boaster, 668 = 2.

Name.	Sex.	Value.	Remainder.	Age (figures indicate years and "A" aged).	Remarks.
Boarder Knight	G.	858	3	3	Knight, 450=9.
Boreen II.	G.	268	7	A.	
Borjulie	G.	251	8	A.	
Borthwick	G.	635	5	4	
Bosh	M.	304	7	A.	
Boss Croker	C.	510	6	3	
Botany Bay	G.	478	1	A.	Bay, 12=3.
Bothways.	G.	436	4	4	
Boucan	H.	79	7	6	
Boulsworth	G.	709	7	5	
Bounce II.	M.	118	1	A.	
Bouquet	F.	38	2	4	
Bourne Bridge	H.	463	4	5	Bridge, 205=7.
Bourton Lass	M.	749	2	5	Lass, 91=1.
Bourtree Hill	G.	853	7	5	Hill, 35=8.
Boveridge	F.	291	3	2	
Bovey Tracy	G.	769	4	4	Tracy, 671=5.
Bow .	G.	8	8	5	
Bowdie Kite	C.	442	1	2	Kite, 420=6.
Bowery	G.	220	4	3	
Bowshot	G.	710	8	4	
Bowsprit	M.	748	1	A.	
Boxer	H.	284	5	6	
Boy Chieftain	G.	557	8	A.	Chieftain, 543=3.
Boy Jim	H.	57	3	6	Jim, 43=7.
Boyne II .	G.	54	9	A.	
Boy of Egremond	H.	432	9	A.	Boy, 14=5.
Braganza .	G.	282	3	A.	
Brain	G.	262	1	5	
Brait	M.	612	9	6	
Bramante	C.	694	1	4	
Brandenburg	G.	529	7	4	
Brandon Bay	M.	319	4	5	Bay, 12=3.
Brandon Head	G.	326	2	6	
Branksome	C.	373	4	4	
Branton Court .	G.	1329	6	5	
Brass Bottle	C.	697	4	2	Bottle, 434=2.
Brasted	G.	677	2	4	
Brauneberg	C.	490	4	2	
Bravo	H.	289	1	A.	
Brayton (By)	C.	662	5	2	
——	C.	662	5	2	
Bread Cutter	H.	836	8	6	
Bread Knife (By)	G.	346	4	3	Knife, 130=4.

Name.	Sex.	Value.	Remainder.	Age (figures indicate years and "A" aged).	Remarks.
Bread Knife (By) .	F.	346	4	2	Knife, 130=4.
—— . . .	F.	346	4	2	Knife, 130=4.
—— . . .	G.	346	4	2	Knife, 130=4.
. . .	C.	346	4	2	Knife, 130=4.
Breadless . .	G.	316	1	6	
Bread Mart . .	C.	821	2	4	Mart, 641=2.
Bread Queen .	F.	296	8	4	Queen, 80=8.
Breakaway .	C.	249	6	3	
Bredenbury .	G.	478	1	4	
Brenda . .	M.	267	6	A.	
Brenda II. .	M.	267	6	A.	
Brer Fox . .	G.	408	1	6	Fox, 192=3.
Brian Born II..	G.	516	3	A.	
Bribery . .	G.	414	9	A.	
Bric a Brac II.	F.	426	3	3	
Brick . .	C.	222	6	4	
Bridegroom .	G.	472	4	A.	
Bridesmaid II..	M.	320	5	6	
Bridge . .	H.	205	7	6	
Bridle . .	G.	236	2	4	
Brigade . .	G.	236	2	A.	
Bright . .	C.	602	8	3	
Bright . .	G.	602	8	4	
Bright Boy . .	G.	616	4	A.	Boy, 14=5.
Bright Gold .	F.	662	5	4	Gold, 60=6.
Bright Grey .	C.	832	4	3	Grey, 230=5.
Brighton II. .	G.	652	4	A.	
Bright Smiles .	F.	792	9	3	Smiles, 190=1.
Brill . . .	C.	232	7	2	
Brillantine .	F.	743	5	4	
Brissac . .	H.	313	7	5	
Brithdir . .	G.	618	6	A.	
Brittle Feet .	M.	1122	6	5	Feet, 490=4.
Broadacre .	C.	439	7	4	
Broad Arrow .	G.	425	2	5	Arrow, 217=1.
Broad Sanctuary	G.	559	1	5	Sanctuary, 351=9.
Broadsword .	H.	478	1	5	Sword, 270=9.
Broadway .	C.	224	8	3	
Broke . .	G.	228	3	2	
Brokenhearted .	F.	898	7	3	
Brom Bones .	C.	309	3	2	
Brookmead .	G.	282	3	5	
Brownberry II.	F.	480	3	2	
Brownberry .	G.	480	3	A.	

Name.	Sex.	Value.	Remainder.	Age (figures indicate years and "A" aged).	Remarks.
Brown Bess II.	M.	330	6	A.	Bess, 72 = 9.
Brown Boy	G.	272	2	A.	Boy, 14 = 5.
Brown Bread	G.	474	6	A.	Bread, 216 = 9.
Brown Ewe	M.	274	4	5	Ewe, 16 = 7.
Brown Hawk	G.	285	6	5	Hawk, 27 = 9.
Brownie	G.	268	7	A.	
Brown Molly	M.	340	7	5	Molly, 82 = 1.
Brown Owl	H.	295	7	5	Owl, 37 = 1.
Brown Princess	M.	718	7	A.	Princess, 460 = 1.
Brown Study	G.	732	3	5	Study, 474 = 6.
Bruise	F.	215	8	4	
Brushton	G.	952	7	2	
Bryn Bras	C.	465	6	3	
Buccaneer (By)	G.	283	4	3	
Bucephalus	G.	242	8	6	
Buck	G.	22	4	3	
Buckingham (By)	F.	137	2	2	
——	G.	137	2	3	
——	F.	137	2	2	
——	F.	137	2	2	
——	C.	137	2	2	
Buckingham Palace	G.	308	2	5	Palace, 171 = 9.
Bucklebury	C.	264	3	2	
Buckram	F.	263	2	2	
Bucksfoot	G.	562	4	A.	
Buck Up	M.	103	4	6	
Budda	F.	7	7	3	
Bude	M.	12	3	5	
Buenos Ayres	C.	399	3	2	
Bugler	C.	252	9	4	
Buglose	F.	65	2	4	
Buller	G.	232	7	4	
Bullet	C.	442	1	3	
Bullfinch	G.	165	3	A.	
Bulrush	C.	532	1	2	
Bumptious (By)	F.	402	6	3	
——	F.	402	6	2	
——	C.	402	6	3	
——	C.	402	6	2	
Bunch of Flowers	M.	514	1	5	Bunch, 55 = 1.
Bunker's Club	C.	384	6	3	Club, 52 = 7.
Bunthorne	G.	709	7	A.	
Burcote	G.	628	7	5	
Burleydam	G.	287	8	A.	

Name.	Sex.	Value.	Remainder.	Age (figures indicate years and "A" aged).	Remarks.
Burmah Ruby . .	C.	461	2	4	Ruby, 218=2.
Burnaby (By) . .	F.	264	3	2	
—— . . .	F.	264	3	3	
Burses . . .	C.	216	9	3	
Burt Hooker . .	C.	833	5	3	Hooker, 231=6.
Burton Bushes. .	G.	961	7	4	Bushes, 309=3.
Burton Pidsea . .	H.	806	5	A.	Pidsea, 154=1.
Busaco . . .	H.	89	8	6	
Busby Stoop . .	M.	620	8	6	Busby, 74=2.
Bushey Belle . .	F.	354	3	2	Belle, 42=6.
Bushey Park (By) .	G.	613	1	2	Park, 301=4.
—— . . .	C.	613	1	3	Park, 301=4.
—— . . .	F.	613	1	3	Park, 301=4.
Bushford Lass . .	M.	677	2	5	Lass, 91=1.
Bushmaster . .	C.	1003	4	2	Master, 701=2.
Bushmills . .	G.	432	9	5	
Business . .	H.	119	2	5	
Busiris . . .	C.	322	7	3	
Butterscotch II. .	G.	687	3	A.	
Butterwort . .	C.	1213	7	3	
By Jove . . .	G.	101	2	A.	
Byzantium . .	M.	510	6	A.	
Cabin Boy (By) .	C.	87	6	2	Boy, 14=5.
—— . . .	C.	87	6	2	Boy, 14=5.
—— . . .	F.	87	6	2	Boy, 14=5.
Cabra . . .	G.	224	8	3	
Cachalot . . .	C.	457	7	3	
Cadran . . .	G.	275	5	4	
Caedmon. . .	C.	126	9	4	
Caerleon . . .	H.	322	7	5	
Cæsar . . .	G.	290	2	6	
Caftan . . .	G.	551	2	5	
Caiman . . .	H.	120	3	6	
Cairn Hill . .	C.	315	9	4	Hill, 35=8.
Cairnryan . .	C.	540	9	3	
Calapita . . .	F.	533	2	4	
Caleb . . .	G.	53	8	4	
Caledon . . .	G.	114	6	A.	
Calendar . . .	M.	305	8	A.	
Calm . . .	F.	62	8	2	
Calypso II. . .	M.	197	8	6	
Camilla . . .	M.	92	2	6	
Campan (By) . .	F.	192	3	2	

Name.	Sex.	Value.	Remainder.	Age (figures indicate years and "A" aged).	Remarks.
Campanone . .	H.	242	8	6	
Candelaria . .	H.	327	3	A.	
Canderos . . .	G.	347	5	5	
Candidate . .	G.	489	3	5	
Candy Tuft . .	M.	965	2	6	Candy, 85=4.
Cannonade . .	F.	135	9	2	
Cannonite . .	M.	527	5	A.	
Canonesse II. . .	M.	197	8	6	
Canter Home . .	G.	722	2	A.	
Cap and Bells II. .	F.	258	6	4	
Caper . . .	G.	301	4	2	
Capitulate . .	F.	947	2	2	
Cappa White . .	G.	430	7	3	Cappa, 24=6.
Capricorn . .	G.	573	6	3	
Capricorn II. . .	M.	573	6	6	
Capresi . . .	M.	381	3	5	
Capstan . . .	H.	611	8	5	
Captain Kettle . .	H.	1011	3	5	Kettle, 460=1.
Captain Laurie . .	G.	793	1	5	Laurie, 242=8.
Captive Pet . .	F.	1071	9	2	Pet, 490=4.
Capucines II. . .	F.	281	2	3	
Carabine . . .	C.	284	5	4	
Caracalla . . .	H.	273	3	5	
Carafe . . .	F.	311	5	2	
Cara Mia . . .	M.	273	3	5	
Carbine (By) . .	C.	283	4	2	
——— . . .	G.	283	4	3	
Carbinia . . .	M.	284	5	5	
Carburton . . .	C.	873	9	3	
Cardington . .	G.	745	7	4	
Caress . . .	M.	283	4	A.	
Carhaix . . .	G.	236	2	A.	
Carholme . . .	G.	272	2	A.	
Carisbrooke . .	F.	509	5	3	
Carle Kemp . .	H.	401	5	5	
Carlocini . . .	G.	377	8	4	
Carlow . . .	G.	257	5	A.	
Carlsbad II. . .	G.	318	3	A.	
Carlton House . .	C.	772	7	2	Carlton, 701=8.
Carnage (By) . .	G.	275	5	4	
Carnethy . . .	G.	696	3	5	
Carnhill . . .	G.	306	9	A.	
Carnis . . .	C.	331	7	3	
Carnmore . . .	G.	517	4	6	

Name.	Sex.	Value.	Remainder.	Age (figures indicate years and "A" aged).	Remarks.
Carnroe . . .	F.	477	9	4	
Caro . . .	C.	227	2	3	
Caro II. . .	C.	227	2	4	
Carolina II. .	M.	308	2	A.	
Carpet Knight .	C.	1161	9	4	Knight, 450=9.
Carrara . .	F.	422	8	4	
Carriden . .	G.	275	5	A.	
Carrier Pigeon .	G.	564	6	6	Pigeon, 133=7.
Carrigavalla .	C.	354	3	4	
Carrigdown .	C.	301	4	4	
Carrignavar .	C.	573	6	2	
Carrington .	G.	741	3	A.	
Carrots . .	M.	681	6	5	
Carsethorn .	G.	948	3	A.	
Carson . .	G.	331	7	6	
Casanova . .	G.	219	3	A.	
Cash Box . .	F.	435	3	3	Box, 114=6.
Cashel . .	C.	351	9	4	
Cashel Boy .	G.	365	5	5	Boy, 14=5.
Casino . .	G.	137	2	4	
Cassine . .	F.	140	5	4	
Cassock's Hope .	M.	252	9	6	Hope, 91=1.
Cassock's Pride .	G.	445	4	A.	Pride, 284=5.
Castasegna .	M.	603	9	A.	
Castellina :	F.	572	5	3	
Castiglione III.	F.	571	4	4	
Castle Belle .	M.	153	9	A.	Belle, 42=6.
Castleblaney (By) .	F.	204	6	3	
Castleconnell .	G.	223	7	A.	
Castle Dance .	C.	226	1	2	Dance, 115=7.
Castle Danger .	H.	369	9	6	Danger, 258=6.
Castlefinn . .	C.	241	7	3	
Castle Hill II. .	M.	146	2	A.	Hill, 35=8.
Castle in Spain .	G.	362	2	A.	Castle, 111=3.
Castle Irwell .	G.	358	7	3	Castle, 111=3.
Castle Ivers .	C.	452	2	4	Castle, 111=3.
Castlenock .	M.	183	3	6	
Castleshira .	C.	612	9	5	
Castlewise .	H.	124	7	5	
Catalpa . .	F.	532	1	3	
Cat Bird . .	F.	627	6	2	
Catcleugh .	F.	487	1	4	
Cateran Lad .	H.	706	4	5	Lad, 35=8.
Cathal . .	G.	456	6	A.	

Name.	Sex.	Value.	Remainder.	Age (figures indicate years and "A" aged).	Remarks.
Catiline . . .	G.	511	7	6	
Catseye II. . .	G.	492	6	A.	
Cauld Blast . .	F.	553	4	2	
Cavaire . . .	G.	311	5	5	
Cavalier II. . .	G.	341	8	A.	
Cavil II. . . .	G.	131	5	A.	
Cawnpore . . .	H.	358	7	5	
Cecilia . . .	F.	171	9	2	
Celebration . .	H.	663	6	5	
Celer . . .	G.	300	3	A.	
Celerity II. . .	G.	710	8	6	
Celibacy . . .	G.	173	2	4	
Celibate . . .	C.	512	8	2	
Cellarman . .	C.	390	3	4	
Celt II. . .	G.	500	5	6	
Cement II. . .	G.	560	2	A.	
Censor . . .	G.	380	2	6	
Centipede . .	G.	614	2	4	
Cento . . .	G.	135	9	4	
Cephaline . .	F.	241	7	2	
Cerdo . . .	C.	270	9	2	
Chacewater . .	M.	680	5	A.	
Chacornac . .	G.	301	4	5	
Chair Fortune .	F.	548	8	2	Chair, 213=6.
Chair of Kildare	G.	558	9	A.	Chair, 213=6.
Chameleon . .	G.	144	9	A.	
Chamois . . .	C.	348	6	2	
Champagne . .	G.	481	4	4	
Chanced Him .	G.	163	1	4	
Chance It . .	M.	515	2	6	
Chancellor . .	C.	344	2	2	
Chant . . .	F.	454	4	3	
Chapeltown . .	M.	570	3	A.	
Chaperon Rouge .	F.	1137	3	3	Chaperon, 631=1.
Chaplin . . .	C.	164	2	4	
Charge . . .	C.	207	9	3	
Charina . . .	M.	255	3	A.	
Charivari . .	G.	495	9	4	
Charles Martel .	C.	975	3	2	Charles, 294=6.
Charlie . . .	G.	244	1	4	
Charlie Dean .	H.	308	2	5	Charlie, 244=1.
Charlton . . .	G.	684	9	5	
Chardon II. . .	C.	258	6	4	
Chartreuse . .	C.	1114	7	3	

Name.	Sex.	Value.	Remainder.	Age (figures indicate years and "A" aged).	Remarks.
Chase . . .	C.	73	1	2	
Chaucer . . .	C.	269	8	2	
Checkman . .	G.	123	6	6	
Cheers . . .	C.	273	3	3	
Cheery Bob . .	G.	229	4	A.	Bob, 6＝6.
Chef . . .	G.	390	3	4	
Cheiro . . .	C.	236	2	3	
Chekoa . . .	H.	40	4	A.	
Chelford . . .	H.	327	3	A.	
Cheltine . . .	F.	503	8	4	
Cheney . . .	G.	73	1	3	
Cheri . . .	C.	223	7	4	
Cheriton Belle . .	F.	705	3	2	Belle, 42＝6.
Cherokee . . .	C.	249	6	4	
Cherry . . .		223	7	5	
Cherry Derry . .	G.	447	6	5	Cherry, 223＝7.
Cherry Laurel . .	C.	495	9	4	Cherry, 223＝7.
Cherry Park . .	F.	524	2	2	Park, 301＝4.
Cherry Pit . .	F.	703	1	2	Pit, 480＝3.
Cherry Ripe (By) .	F.	503	8	2	Cherry, 223＝7.
—— . . .	C.	503	8	2	Cherry, 223＝7.
—— . . .	F.	503	8	3	Cherry, 223＝7.
—— . . .	C.	503	8	3	Cherry, 223＝7.
Cherry Wife . .	M.	309	3	5	Wife, 86＝5.
Cherub . . .	G.	215	8	5	
Cheshire . . .	H.	513	9	A.	
Chess . . .	M.	73	1	5	
Chester . . .	G.	673	7	A.	
Chesterfield . .	G.	797	5	A.	
Chevalier II. . .	G.	631	1	4	
Cheviot . . .	C.	503	8	3	
Chevrenil . . .	F.	339	6	3	
Chevy Chase . .	G.	176	5	A.	Chase, 73＝1.
Chez Moi . . .	M.	357	6	5	
Chiana . . .	C.	65	2	2	
Chicane . . .	G.	83	2	4	
Chief . . .	G.	93	3	A.	
Chilblair . . .	M.	125	8	5	
Childless . . .	C.	137	2	4	
Child Princess . .	F.	497	2	2	Princess, 460＝1.
Child's Guide . .	C.	131	5	4	Guide, 34＝7.
Child Waters . .	H.	704	2	5	
Childwick (By) .	F.	57	3	2	
—— . . .	C.	57	3	2	

Name.	Sex.	Value.	Remainder.	Age (figures indicate years and "A" aged).	Remarks.
Childwickbury . .	G.	269	8	6	
Childwit . . .	G.	443	2	4	
Chilgrove . . .	M.	339	6	5	
Chillagoe . . .	H.	60	6	5	
Chillingworth . .	G.	714	3	A.	
Chillon . . .	C.	94	4	3	
Chilpiquin . .	F.	183	3	2	
China Bead . .	F.	70	7	2	Bede, 16=7.
Chinewood . .	C.	63	9	4	
Chionia . . .	M.	80	8	A.	
Chips . . .	G.	143	8	A.	
Chirtle . . .	F.	633	3	3	
Chiseldom . .	G.	147	3	6	
Chiselhampton .	G.	589	4	A.	
Chit Chat . .	G.	807	6	A.	
Chittabob (By) .	C.	410	5	2	
—— . . .	F.	410	5	2	
Chitty . . .	F.	413	8	4	
Chocolate . .	M.	455	5	A.	
Chocolate II. .		455	5	6	
Choctaw . . .	G.	426	3	4	
Choir Girl . .	M.	487	1	5	Girl, 250=7.
Chon Kina . .	M.	126	9	6	
Chon Kina . .	G.	126	9	A.	
Chop . . .	G.	85	4	A.	
Chouette . . .	F.	419	5	2	
Chrysis . . .	F.	350	8	2	
Chump . . .	C.	43	7	4	
Chupatty . . .	F.	494	8	3	
Cigarette . . .	F.	690	6	2	
Cincinnatus . .	G.	741	3	A.	
C. I. V. . . .	H.	171	9	5	
Circassian (By) .	G.	641	2	2	
Cissy . . .	M.	130	4	5	
Clairetta . . .	F.	671	5	2	
Clamour . . .	F.	291	3	2	
Clampin . . .	C.	221	5	3	
Clandon Lad . .	G.	136	1	2	
Clansman II. .	G.	251	8	A.	
Claqueur . . .	G.	271	1	4	
Clare Man . .	G.	351	9	A.	
Claremorris . .	G.	562	4	6	
Clarendon Road .	F.	365	5	4	Road, 210=3.
Claret . . .	G.	651	3	3	

Name.	Sex.	Value.	Remainder.	Age (figures indicate years and "A" aged).	Remarks.
Claribel . . .	F.	293	5	2	
Clarine . . .	F.	311	5	3	
Clarissa . . .		312	6	5	
Clarnico . . .	F.	327	3	4	
Claudia . . .	M.	67	4	A.	
Clean Gone . .	G.	182	2	6	
Clear Note . .	M.	716	5	5	
Cleator Moor .	H.	906	6	6	Cleator, 660 = 3.
Cleevethorpe .	H.	827	8	5	
Cleg Kelly . .	G.	150	6	4	Cleg, 80 = 8.
Cleo		66	3	3	
Cleopatra . .	F.	748	1	2	
Clerval . . .	H.	371	2	6	
Cliff	C.	130	4	3	
Cliftonhall . .	C.	617	5	3	
Clinker II. . .	G.	320	5	A.	
Clisiade . . .	F.	124	7	5	
Clockwork . .	F.	298	1	2	
Cloister III. .	G.	712	1	A.	
Clonakilty Bay	C.	551	2	2	Bay, 14 = 5.
Clonard . . .	G.	311	5	4	
Clondalkin . .	H.	207	9	A.	
Clonsilla II. .	G.	193	4	A.	
Clontra . . .	M.	703	1	5	
Clorania . . .		318	3	2	
Closure . . .	H.	263	2	A.	
Cloten . . .	G.	516	3	2	
Cloture . . .	C.	259	7	2	
Cloud . . .	G.	60	6	6	
Cloven Foot .	G.	666	9	5	
Cloverley . .	C.	376	7	4	
Club Bail . .	F.	94	4	2	Bail, 42 = 6.
Club Force . .	G.	398	2	5	Force, 346 = 4.
Club Gossip .	F.	534	3	4	Gossip, 162 = 9.
Clwyd (By) .	F.	60	6	2	
—— . . .	F.	60	6	2	
Clwyd II. . .	G.	60	6	2	
Clynder . . .	F.	304	7	3	
Coal Sack . .	C.	137	2	3	Sack, 81 = 9.
Coal Tax . .	G.	567	9	4	Tax, 511 = 7.
Cobbler II. .	G.	254	9	6	
Cobden . . .	G.	78	6	5	
Cobweb . . .	C.	46	1	3	
Cochlearia . .	M.	284	5	5	

Name.	Sex.	Value.	Remainder.	Age (figures indicate years and "A." aged).	Remarks.
Cockenheugh . .	G.	113	5	4	
Cockhill . . .	C.	77	5	4	
Cock Robin . .	G.	296	8	A.	Robin, 254 = 2.
Cocotte . . .	F.	448	7	4	
Codford . . .	C.	310	4	4	
Codoman . . .	C.	122	5	5	
Coenraad . . .	G.	292	4	A.	
Cogia . . .	G.	40	4	2	
Cohiltown . .	G.	517	4	6	
Coincidence . .	F.	251	8	2	
Coiner . . .	G.	272	2	4	
Colchester . .	C.	729	9	2	
Coldra . . .	C.	261	9	3	
Colette II. . .	M.	462	3	6	
Colleen Bawn . .	F.	204	6	4	
Colleen Das Dhoun .	M.	241		5	
College Queen . .	F.	145	1	4	Queen, 80 = 8.
Colleger . . .	G.	265	4	5	
Colon . . .	G.	106	7	6	
Colonel Bartlett .		1343	2	A.	Bartlett, 1043 = 8.
Colonel Wozac . .	C.	334	1	3	Wozac, 34 = 7.
Columbary . .	G.	308	2	4	
Columbine II. . .	M.	158	5	A.	
Combat . . .	F.	463	4	2	
Combe Martin . .	G.	757	1	5	Combe, 66 = 3.
Comber . . .	G.	268	7	4	
Come Again II. .	M.	141	6	6	
Comet III. . .	G.	472	4	A.	
Come to Order . .	G.	872	8	6	
Commandeer . .	G.	316	1	3	
Commodore . .	C.	278	8	4	
Common (By) . .	F.	112	4	2	
Communist . .	G.	586	1	3	
Competent . .	F.	1002	3	4	
Compliment . .	G.	672	6	4	
Composer . .	G.	353	2	4	
Composite . .	G.	547	7	4	
Compton Lad . .	G.	547	7	4	Lad, 35 = 8.
Comrade II. . .	H.	276	6	A.	
Conejo . . .	G.	95	5	4	
Confectioner . .	G.	730	1	6	
Conform . . .	G.	392	5	4	
Congratulation . .	F.	677	2	4	
Congratulation . .	M.	677	2	A.	

Name.	Sex.	Value.	Remainder.	Age (figures indicate years and " A " aged).	Remarks.
Coningsby . . .	C.	214	7	4	
Conna . . .	G.	73	1	A.	
Connaught King .	G.	563	5	A.	
Con O'Ryan . .	G.	339	6	5	
Conquering Hero .	G.	583	7	A.	Hero, 221=5.
Conqueror II. . .	G.	492	6	A.	
Consolation . .	G.	519	6	5	
Consoler . . .	F.	366	6	4	
Consolida. . .	F.	167	5	2	
Conspiracy . .	M.	481	4	5	
Constant . . .	F.	982	1	3	
Consternation . .	G.	1133	8	5	
Contraband . .	M.	730	1	5	
Convamore . .	C.	399	3	4	
Conventicle . .	F.	662	5	2	
Convent Maid . .	M.	666	9	6	Maid, 54=9.
Convert . . .	G.	752	5	A.	
Coo-ee . . .	F.	37	1	3	
Cool Assurance .	C.	673	7	2	Assurance, 617=5.
Coolattin . . .	C.	507	3	2	
Coolgardie . .	G.	291	3	A.	
Coolock . . .	C.	78	6	4	
Cookham . . .	G.	85	4	4	
Coon . . .	C.	76	4	2	
Copyright. . .	F.	712	1	3	
Coragh Melton. .	F.	776	2	2	Melton, 530=8.
Coralite . . .	M.	657	9	5	
Coral Sea . . .	M.	326	2	5	Sea, 70=7.
Corbal Lis . .	G.	348	6	A.	
Corbella . . .	F.	269	8	2	
Corie Lynn . .	M.	316	1	5	
Cork . . .	G.	242	8	A.	
Cormac . . .	C.	283	4	2	
Corner . . .	G.	472	4	A.	
Corner Boy . .	C.	486	9	3	Boy, 14=5.
Corn Exchange .	G.	420	6	6	Exchange, 148=4.
Cornfield . . .	G.	396	9	A.	
Corn Flag . .	F.	403	7	4	Flag, 131=5.
Cornflour . . .	G.	588	3	A.	
Cornice . . .	H.	332	8	6	
Corn King . .	G.	362	2	5	
Cornshifter . .	C.	1152	9	2	
Coronado . . .	C.	289	1	3	
Coronea . . .	F.	289	1	3	

Name.	Sex.	Value.	Remainder.	Age (figures indicate years and "A" aged).	Remarks.
Coronet . . .	G.	688	4	6	
Coronet II. . .	G.	688	4	A.	
Coroun . . .	G.	278	8	4	
Corunna . . .	F.	279	9	3	
Cosmopolite . .	G.	610	7	A.	
Cossack . . .	C.	102	3	4	
Cossack Post . .	G.	648	9	4	Post, 546=6.
Cosy Nook . .	H.	119	2	6	Nook, 76=4.
Cotherstone . .	G.	1143	9	6	
Cottage Maid . .	F.	479	2	3	Maid, 54=9.
Cottager . . .	C.	635	5	4	
Cottenshope . .	G.	858	3	6	
Cottesmore II.. .	G.	738	9	A.	
Coucou . . .	C.	52	7	4	
Cough Lozenge .	H.	387	9	A.	
Coulin . . .	M.	106	7	5	
Councillor . .	G.	366	6	A.	
Counsellor II. . .	G.	366	6	A.	
Council of Trent .	C.	1307	2	4	Council, 166=4.
Countermark . .	C.	937	1	2	
Countess II. . .	M.	546	6	A.	
Countess II. . .	M.	546	6	6	
Countess IV. . .	M.	546	6	5	
Countess Helena .	F.	652	4	3	Helena, 106=7.
Countess Hermit .	M.	1191	3	5	Hermit, 645=6.
Country Boy . .	C.	694	1	2	Boy, 14=5.
Country Girl . .	F.	930	3	2	Girl, 250=7.
County Life . .	G.	596	2	6	Life, 110=2.
Coup Double . .	C.	142	7	3	
Coureur des Bois .	C.	260	8	4	Coureur, 237=3.
Courlan . . .	G.	307	1	5	
Couronnement . .	G.	368	8	3	
Court Amour . .	F.	873	9	4	Amour, 247=4.
Court Belle . .	F.	668	2	4	Belle, 42=6.
Courtier II. . .	G.	836	8	A.	
Cove . . .	C.	106	7	3	
Covent Garden .	G.	837	9	A.	Garden, 275=5.
Covert Hack . .	G.	728	8	A.	Hack, 26=8.
Cowley . . .	C.	66	3	3	
Crackers . . .	M.	501	6	6	
Cracky . . .	G.	251	8	6	
Craddoxtown . .	G.	761	5	A.	
Crafton (By) . .	C.	751	4	4	
——— . . .	C.	751	4	2	

Name.	Sex.	Value.	Remainder.	Age (figures indicate years and "A" aged).	Remarks.
Crafton (By) . .	G.	751	4	3	
Crafty Agnes . .	F.	852	6	3	Agnes, 141 = 6.
Crafty Party . .	C.	1402	7	4	Party, 691 = 7.
Crafty Thought .	G.	1518	6	5	Thought, 807 = 6.
Cragielea . . .	M.	275	5	A.	
Cragsman . .	G.	391	4	4	
Craig Ard. . .	G.	456	6	6	
Craigengillan . .	F.	411	6	4	
Craigflower . .	F.	568	1	3	
Craig Graham . .	H.	516	3	5	Craig, 250 = 7.
Craighampton . .	F.	746	8	2	
Craig Royston (By) .	G.	972	9	3	Craig, 250 = 7.
Crarae . . .	H.	431	8	5	
Crasher . . .	G.	721	1	A.	
Crautacaun . .	G.	703	1	4	
Crawley Lass . .	F.	353	2	4	Lass, 91 = 1.
Cream . . .	F.	270	9	4	
Creditor . . .	C.	834	6	3	
Creepy Crawley .	G.	582	6	A.	Crawley, 262 = 1.
Creme de Menthe .	G.	375	6	A.	Cream, 270 = 9.
Cremore . . .	F.	476	8	4	
Creolin . . .	G.	316	1	5	
Crevette . . .	F.	720	9	4	
Crimson Heather .	F.	589	4	2	Heather, 219 = 3.
Crispus . . .	H.	420	6	A.	
Cromwell . . .	G.	308	2	A.	
Cromwellsgate . .	G.	798	6	5	
Cronborg . . .	G.	500	5	A.	
Crookhaven . .	F.	382	4	2	
Croom . . .	G.	266	5	4	
Croppy Boy . .	G.	248	5	A.	Boy, 14 = 5.
Croquet . . .	M.	256	4	6	
Crossburn . .	G.	534	3	A.	
Cross Purpose . .	M.	649	1	A.	
Crowle . . .	C.	256	4	3	
Crown . . .	G.	276	6	A.	
Crown Derby . .	F.	492	6	3	
Crown Equerry .	H.	527	5	5	Equerry, 251 = 8.
Crown Imperial .	G.	647	8	5	
Crow Needle . .	F.	320	5	2	
Crownet . . .	G.	686	2	4	
Crow Not . .	F.	678	3	5	
Crowood . . .	M.	236	2	A.	
Crow's Nest . .	G.	753	6	6	

Name.	Sex.	Value.	Remainder.	Age (figures indicate years and "A" aged).	Remarks.
Croyland . . .	G.	317	2	2	
Crumbs . . .	M.	322	7	A.	
Crystal Palace . .	G.	881	8	A.	
Csardas . . .	C.	325	1	3	
Culverin . . .	H.	380	2	6	
Cumberland . .	G.	347	5	5	
Cunninghame . .	G.	185	5	A.	
Cupbearer . . .	C.	512	8	3	
Curate . . .	G.	646	7	5	
Curlew . . .	F.	256	4	4	
Curls . . .	F.	310	4	4	
Curragh Belle . .	F.	283	4	4	
Curraghmore . .	C.	487	1	4	
Curtain . . .	F.	670	4	2	
Cush . . .	F.	320	5	4	
Cushat Doo . .	F.	731	2	2	
Cushendun . .	G.	434	2	A.	
Cut and Come Again	C.	616	4	3	
Cutaway . . .	G.	437	5	5	
Cutler . . .	H.	650	2	A.	
Cyclonic . . .	F.	186	6	4	
Cynosurus . .	G.	432	9	A.	
Cyrene Hazel . .	M.	363	3	5	Hazel, 43=7.
Cyrlena . . .	F.	351	9	4	
Czigany . . .	F.	88	7	2	
Dalbernon . .	C.	307	1	3	
Dai Nippon . .	G.	119	2	5	Dai, 15=6.
Dainty Dish . .	M.	778	4	A.	
Dairy Maid . .	F.	278	8	3	Maid, 54=9.
Daisy II. . . .	M.	31	4	A.	
Daisy's Joy . .	G.	53	8	A.	Joy, 15=6.
Daka II. . . .	C.	26	8	2	
Dakota III. . .	C.	432	9	3	
Dalbeattie . .	C.	457	7	4	
Daldorch . . .	G.	261	9	6	
Dalemore . .	G.	290	2	A.	
Dalesman . .	C.	194	5	3	
Dalhousie . .	G.	62	8	4	
Dalmeny . . .	C.	145	1	3	
Dalmorton . .	G.	727	7	5	
Dam . . .	G.	45	9	4	
Dam II. . . .	G.	45	9	A.	
Damaraland . .	M.	332	8	5	

Name.	Sex.	Value.	Remainder.	Age (figures indicate years and "A" aged).	Remarks.
Dame d'Or . .	F.	260	8	3	
Damersol . . .	F.	341	8	4	
Dancer . . .	H.	315	9	A.	
Dancer II. . .	M.	315	9	5	
Dancing Gal . .	M.	236	2	A.	Gal, 51 = 6.
Dancing Laddie II. .	G.	230	5	A.	Laddie, 45 = 9.
Dancing Master .	G.	886	4	A.	Master, 701 = 8.
Dancing Nun . .	F.	285	6	2	
Dandolo . . .	C.	101	2	4	
Dandy Boy . .	C.	83	2	3	Boy, 14 = 5.
Dandy Fifth . .	H.	634	4	6	Dandy, 69 = 6.
Dandy Lad . .	C.	104	5	5	Dandy, 69 = 6.
Danger . . .	M.	258	6	5	
Dangerous . .	F.	318	3	4	
Dangle . . .	G.	105	6	6	
Daphnis . . .	H.	195	6	A.	
Dapper . . .	F.	207	9	2	
Dare Devil II. . .	G.	338	5	A.	Devil, 124 = 7.
Darius . . .	C.	265	4	3	
Dark David . .	H.	240	6	6	David, = .
Dark Eye . .	M.	236	2	5	Eye, 11 = 2.
Dark Magic . .	F.	289	1	4	Magic, 64 = 1.
Dark 'Un . . .	M.	276	6	A.	
Darling Clara . .	M.	557	8	A.	Clara, 252 = 9.
Darling Nancy .	M.	506	2	6	Fancy, 201 = 3.
Darnel . . .	F.	295	7	4	
Darnley (By) . .	G.	295	7	4	
—— . . .	G.	295	7	2	
Darraghmore . .	G.	471	3	A.	
D'Artagnan . .	G.	726	6	5	
Dash . . .	G.	305	8	2	
Dathi . . .	G.	420	6	5	
Datura II. . .	C.	612	9	3	
Dauntless II. . .	F.	556	7	4	
Dave . . .	G.	94	4	4	
David II. . . .	H.	15	6	A.	
David Devant . .	C.	160	7	3	David, 15 = 6.
David Grieve . .	C.	325	1	4	David, 15 = 6.
David Harum . .	G.	261	9	5	David, 15 = 6.
Day Lily . . .	F.	84	3	3	Lily, 70 = 7.
Dazzle (By) . .	C.	42	6	2	
Dead Certainty .	F.	1138	4	2	
Dearslayer . .	G.	514	1	6	
Death Duty . .	M.	849	3	A.	Duty, 430 = 7.

Name.	Sex.	Value.	Remainder.	Age (figures indicate years and "A" aged).	Remarks.
Debacle . . .	F.	67	4	3	
Debutante . .	F.	867	3	2	
Decave . . .	G.	124	7	3	
Deemarion . .	C.	315	9	4	
Deemster (By) . .	C.	714	3	3	
—— . . .	F.	714	3	2	
Deepdene . .	G.	158	5	A.	
Deep Level . .	G.	244	1	A.	
Deep Sea . .	G.	164	2	A.	
Deep Thoughts .	M.	901	1	A.	
Deesartagh . .	H.	623	2	A.	
Defence . . .	G.	214	7	A.	
Defosse . . .	F.	156	3	2	
Degenerate . .	G.	677	2	A.	
Deianira . . .	F.	266	5	3	
De'il . . .	G.	44	8	3	
De Lacy . . .	H.	115	7	5	Lacy, 101 = 2.
Delarey . . .	G.	255	3	5	
Delaware . . .	M.	261	9	A.	
Delivery . . .	G.	334	1	6	
Delphic . . .	G.	144	9	4	
Delta . . .		445	4	A.	
Demon . . .	G.	104	5	5	
Demonship . .	M.	484	7	A.	
Denis Richard . .	C.	531	9	3	
Departed . . .	C.	699	6	3	
Deportment . .	H.	1200	3	5	
Dermot Asthore .	H.	1316	2	A.	Dermot, 644 = 5.
De Rougemont .	G.	611	8	A.	
Derryquin . .	C.	294	6	4	
Descender . .	H.	338	5	6	
Deseo . . .	F.	37	1	4	
Desert Chief . .	G.	714	3	4	Chief, 93 = 3.
Desert Flower . .	M.	939	3	A.	Flower, 318 = 3.
Designation . .	C.	442	1	4	
Desinvolture . .	F.	386	8	3	
Despair (By) . .	F.	364	4	4	
—— . . .	G.	364	4	2	
—— . . .	C.	364	4	4	
—— . . .	C.	364	4	2	
—— . . .	C.	364	4	2	
—— . . .	G.	364	4	2	
—— . . .	G.	364	4	2	
—— . . .	C.	364	4	3	

Name.	Sex.	Value.	Remainder.	Age (figures indicate years and "A." aged).	Remarks.
Despair Not	C.	816	6	3	
Dessechee	F.	97	7	3	
Destroyer	H.	886	4	5	
Destroyer II.	G.	886	4	A.	
Detail	G.	454	4	6	
Detour	G.	620	8	4	
Deuce of a Daisy	M.	203	5	6	Deuce, 80=8.
Deuce of Clubs (By)	F.	273	3	2	
———	F.	273	3	2	
———	F.	273	3	2	
———	F.	273	3	2	
———	G.	273	2	2	
Deva	F.	95	5	3	
Devilet	G.	534	3	A.	
Devilkin	F.	194	5	4	
Devil's Dance	C.	299	2	4	Dance, 115=7.
Devote	M.	500	5	3	
Dewberry	M.	242	8	A.	
De Wet II.	G.	430	7	5	
De Wet III.	G.	430	7	A.	
De Wet IV.	G.	430	7	6	
Dewi Sant	C.	541	1	5	Dewy, 30=3.
Diarmaid	C.	265	4	3	
Dickford		308	2	4	
Dick Turpin II.	G.	754	7	5	Dick, 24=6.
Dick Wynne	G.	80	8	4	Dick, 24=6.
Dictature	C.	628	7	4	
Didn't Know	C.	514	1	4	
Dido	M.	14	5	A.	
Dinah	F.	59	5	3	
Dinna Greet	C.	685	1	4	Dinna, 55=1.
Dinna Ken	H.	135	9	5	Dinna, 55=1.
Diplomat (By)	F.	561	3	2	
Dippenhall	G.	103	4	5	
Dirkhampton	G.	720	9	A.	
Disarmed	G.	310	4	6	
Disdainful	M.	238	4	5	
Dismay	G.	114	6	6	
Dispensation	C.	615	3	2	
Distressful	G.	844	7	6	
Disturbance II.	G.	777	3	A.	
Divination	F.	485	8	3	
Dixie	C.	124	7	4	
Dixie's Land	C.	216	9	4	Land, 85=4.

Name.	Sex.	Value.	Remainder.	Age (figures indicate years and "A" aged).	Remarks.
Dobbins (By) . .	F.	118	1	2	
Doctor . . .	G.	626	5	A.	
Dr. Barnardo . .	C.	1090	1	3	Barnardo, 464 = 5.
Dr. Charlie . .	G.	870	6	5	Charlie, 244 = 1.
Dr. Jim . . .	G.	669	3	6	Jim, 43 = 7.
Doctrine . . .	F.	676	1	3	
Doddington . .	H.	530	8	6	
Dodor . . .	G.	214	7	A.	
Dodragh . . .	F.	211	4	4	
Dodridge . . .	G.	213	6	3	
Dog Rose (By) .	F.	239	5	2	Rose, 213 = 6.
Doleful . . .	G.	150	6	5	
Dollar III. . .	G.	236	2	5	
Dolly Queen . .	F.	126	9	3	Dolly, 46 = 1.
Dolphin III. . .	G.	166	4	A.	
Domineer . .	H.	306	9	5	
Dominie . . .	G.	106	7	6	
Dominie II. . .	H.	106	7	6	
Domremy . . .	C.	306	9	2	
Donah . . .	F.	65	2	3	
Donalbane . .		148	4	A.	
Donatello. . .	C.	507	3	2	
Donative . . .	C.	541	1	3	
Donauvellen . .	F.	231	6	4	
Don Conquest . .	G.	618	6	4	Conquest, 562 = 4.
Donegan . . .	G.	140	5	A.	
Donizetti . . .	G.	487	1	5	
Donna Lorna . .	F.	348	6	4	Lorna, 287 = 8.
Donnybrook Fair .	C.	598	4	3	
Donovan (By) . .	F.	186	6	2	
Don Quixote . .	C.	505	1	4	Quixote, 449 = 8.
Donzelle . . .	F.	107	8	2	
Doochary . . .	G.	240	6	3	
Dora Dora . .	F.	422	8	3	
Doric II. . .	G.	226	1	A.	
Doricles . . .	C.	273	3	4	
Dorine . . .	F.	270	9	3	
Dorothy Lee . .	F.	667	1	4	Dorothy, 627 = 6.
Dorothy Melton .	F.	1157	5	3	Dorothy, 627 = 6.
Dorothy W. . .	F.	679	4	2	Dorothy, 627 = 6.
Dorris Dear . .	M.	480	3	5	
D'Orsay . . .	C.	276	6	2	
Double Dealer . .	C.	280	1	3	Dealer, 244 = 1.
Doubledutch . .	G.	43	7	3	Dutch, 7.

Name.	Sex.	Value.	Remainder.	Age (figures indicate years and "A", aged).	Remarks.
Doubloon (By) . .	C.	92	2	4	
Doubtful II. . .	M.	520	7	5	
Doubtful Honour .	F.	772	7	4	Honour, 252=9.
Douceur . . .	C.	280	1	2	
Doux Pays II. . .	C.	107	8	4	
Dover Bay . .	F.	302	5	3	Bay, 12=3.
Dovercourt . .	G.	916	7	5	
Dowf . . .	H.	90	9	A.	
Downham . .	H.	105	6	5	
Dragon Fly . .	G.	395	8	5	Fly, 120=3.
Drake . . .	G.	234	9	A.	
Dramatist . .	G.	1106	8	5	
Draper . . .	G.	494	8	A.	
Drayton Victor .	G.	1364	5	6	Victor, 700=7.
Dreadnought II. .	G.	670	4	A.	
Dreyfus . . .	H.	354	3	5	
Driftwood . .	C.	694	1	3	
Driver . . .	G.	484	7	A.	
Drogheda . .	G.	245	2	A.	
Dromara . .	M.	455	4	6	
Dromonby Yet. .	M.	724	4	A.	
Druid . . .	G.	214	7	A.	
Druidical . .	G.	264	3	4	
Druid's Mark .	C.	482	5	2	Mark, 261=9.
Drumbrackan .	C.	517	4	3	
Drumcree . .	G.	474	6	A.	
Drumdowney .	G.	394	7	5	
Drumree . . .	G.	454	4	6	
Drumeel . .	C.	284	5	3	
Drumshoreland .	C.	835	7	2	
Dry Still . .	F.	704	2	2	
Duart Bay . .	C.	623	2	2	
Duchess of Kent .	M.	638	8	A.	Duchess, 77=5.
Duckey . .	G.	34	7	6	
Duck Hawk . .	F.	51	6	3	Hawk, 27=9.
Duel II. . .	G.	44	8	5	
Duenna . .	M.	65	2	5	
Duhallow . .	G.	46	1	A.	
Duke of Westminster	C.	1148	5	3	Duke, 40=4.
Duke's Seal .	G.	200	2	A.	Seal, 100=1.
Duke's Token .	F.	576	9	3	Token, 476=8.
Duke William .	C.	126	9	3	William, 86=5.
Dula . . .	F.	41	5	4	
Dulwich (By) .	F.	37	1	3	

Name.	Sex.	Value.	Remainder.	Age (figures indicate years and "A" aged).	Remarks.
Dumps	G.	184	4	2	
Dunboy	G.	68	5	4	
Duncan	C.	124	7	3	
Dundas II.	G.	119	2	A.	
Dundonald	C.	144	9	4	
Dunkettle		514	1	6	
Dunlavin	H.	215	8	5	
Dunlin	G.	134	8	3	
Dunmurry	C.	304	7	3	
Dunnabie	G.	67	4	5	
Dunston	C.	564	6	3	
Durrington	H.	724	4	A.	
Durrow	G.	210	3	4	
Dusky Lad	C.	129	3	2	Lad, 35 = 8.
Dutch Bill	G.	39	3	A.	Bill, 32 = 5.
Dutch Bride	F.	213	6	2	Bride, 206 = 8.
Dynamite II.	G.	505	1	A.	
Eagle's Visit	C.	608	5	2	Visit, 487 = 1.
Eaglet	G.	471	3	3	
Earl	G.	231	6	6	
Earl's Seat	F.	761	5	2	Seat, 470 = 2.
Early Purl	M.	551	2	5	Purl, 310 = 4.
Earthstopper	G.	1270	1	3	
Earthworm	G.	852	6	A.	
East Anglican	G.	652	4	A.	Anglican, 181 = 1.
Easter Eve	M.	762	6	5	Eve, 91 = 1.
Easter Gift II.	H.	1171	1	A.	Gift, 500 = 5.
Eastern Friars	G.	1262	2	5	Friars, 541 = 1.
Eastern Light	H.	1151	8	6	Light, 430 = 7.
Easter Ogue	G.	698	5	A.	Ogue, 27 = 9.
Easter Prize	C.	958	4	4	Prize, 287 = 8.
Eastertide	M.	1075	4	5	
Eastleigh	G.	511	7	A.	
Eastward	G.	682	7	4	
Eatherbloom	M.	694	1	5	
Eau de Vie	G.	111	3	A.	
Ebor	G.	213	6	A.	
Eboracum	G.	280	1	6	
Ecclefechan II.	G.	222	6	5	
Echelon	F.	96	6	4	
Echo	C.	37	1	3	
Echo II.	G.	37	1	4	
Echo III.	G.	37	1	A.	

Name.	Sex.	Value.	Remainder.	Age (figures indicate years and "A" aged).	Remarks.
Ecila . . .	F.	112	4	3	
Eclipse . . .	H.	201	3	A.	
Economist . .	C.	573	6	2	
Ecstacy II. . .	G.	552	3	6	
Ecton . . .	G.	481	4	5	
Eddisbury . .	G.	287	8	4	
Edmee . . .	M.	65	2	6	
Egmont . . .	H.	123	6	5	
Ego	H.	37	1	A.	
Egyptian Briar .	G.	847	1	4	Briar, 403=7.
Eiderdown . .	F.	275	5	3	
Eight o'Clock .	F.	490	4	4	
Eileen Allanah II. .	M.	188	8	A.	Eileen, 101=2.
Eileen Rose . .	F.	314	8	2	Eileen, 101=2.
Eileen Violet .	F.	637	7	4	Eileen, 101=2.
Elands Laagte .	C.	618	6	4	Laagte, 462=3.
Elastic . .	M.	512	8	6	
El Bodon . .	G.	103	4	A.	
Eldest Son .	G.	625	4	2	Son, 110=2.
Electric Current .	H.	1351	1	6	Current, 670=4.
Elena . . .	M.	92	2	6	
Elephantine .	F.	642	3	3	
Elfdale . .	G.	165	3	4	
Elfeet . .	C.	531	9	2	
Elfrida . .	M.	326	2	A.	
Elizabeth . .	F.	456	6	4	
Ella Crag .	F.	283	4	4	Ella, 42=6.
Ellaline II. .	M.	132	6	6	
Elleborin . .	F.	309	3	2	
Ellen . . .	M.	91	1	5	
Ellerslie . .	F.	341	8	3	
Elton . . .	G.	491	5	A.	
Elunilla . .	M.	128	2	5	
Emerald II. .	G.	295	7	A.	
Emerald Gem .	F.	348	6	4	Gem, 53=8.
Emerald Star .	G.	956	2	6	Star, 661=4.
Emeraud . .	F.	267	6	3	
Emigrant .	M.	722	2	5	
Emigrant II. .	G.	722	2	4	
Emily Melton .	F.	621	9	5	Emily, 91=1.
Emir . . .	C.	251	8	4	
Emissary .	G.	321	6	3	
Emperador .	C.	542	2	4	
Empress II. .	M.	401	5	5	

8

Name.	Sex.	Value.	Remainder.	Age (figures indicate years and "A" aged).	Remarks.
Empressement . .	G.	901	1	5	
Encombe . . .	C.	121	4	4	
Encore . . .	G.	287	8	4	
Endeavour . .	C.	345	3	2	
Engineer . . .	C.	314	8	4	
Engineer II. . .	G.	314	8	A.	
England's Queen .	F.	295	7	3	Queen, 80 = 8.
Enniskerry . .	G.	361	1	A.	
Eno	G.	67	4	4	
Enoch Arden . .	G.	337	4	5	
Enslin . . .	F.	201	3	3	
Enter . . .	F.	661	4	3	
Enterprise (By) .	G.	948	3	2	
———— . . .	G.	948	3	3	
Entertainer . .	G.	1121	5	4	
Enthusiast (By) .	G.	950	5	2	
Eon	C.	61	7	2	
Eoos . . .	G.	77	5	A.	
Epsom Lad . .	H.	226	1	5	Lad, 35 = 8.
Equator . . .	G.	637	7	4	
Equerry . . .	G.	251	8	6	
Eraquois . . .	G.	239	5	3	
Erbin . . .	G.	253	1	A.	
Eremite . . .	G.	661	4	5	
Ergot . . .	C.	227	2	3	
Eric	G.	231	6	A.	
Erik	H.	231	6	6	
Ermyntrude . .	F.	901	1	3	
Errol . . .	G.	243	9	3	
Escalonia . . .	F.	189	9	3	
Escarpe . . .	H.	372	3	5	
Escroqueur . .	C.	527	5	3	
Esme Lee . .	M.	161	8	6	
Esmeralda II . . .	M.	357	6	5	
Essex . . .	H.	161	8	A.	
Estar . . .	G.	671	5	A.	
Esther Waters . .	M.	1343	2	A.	Waters, 667 = 1.
Estonia . . .	F.	538	7	3	
Estula . . .	C.	502	7	3	
Eteocles . . .	G.	494	8	6	
Eternity . . .	G.	1061	8	4	
Ethelbruce . .	C.	714	3	3	
Ethel May . .	F.	496	1	2	
Ethel Norah . .	M.	703	1	6	Ethel, 446 = 5.

Name.	Sex.	Value.	Remainder.	Age (figures indicate years and "A" aged).	Remarks.
Ethelred . . .	H.	660	3	6	
Ethel's Darling .	M.	811	1	5	Darling, 305 = 8.
Ethel Star . .	G.	1107	9	A.	Ethel, 446 = 5.
Ethelwulf . . .	C.	562	4	4	
Ethna . . .	F.	467	8	4	
Etoilee . . .	F.	443	2	4	
Etruscan . . .	C.	741	3	3	
Ettrick . . .	H.	631	1	6	
Eugene Stratton .	H.	1190	2	6	Eugene, 79 = 7.
Eugie . . .	M.	29	2	5	
Euonymus . .	M.	176	5	5	
Euphrasia . .	M.	362	2	6	
Evan . . .	C.	141	6	3	
Ever Constant . .	M.	1273	4	5	Constant, 982 = 1.
Everleigh . . .	G.	331	7	A.	
Evreux . . .	C.	304	7	3	
Exaltation . .	F.	820	1	4	
Excepcional . .	H.	321	6	A.	
Exchange . .	F.	148	4	4	
Exchequer . .	C.	324	9	2	
Exedo . . .	C.	111	3	4	
Exema . . .	C.	89	8	3	
Exempler . . .	G.	398	2	6	
Exile . . .	G.	78	6	A.	
Exile II. (By) . .	F.	78	5	3	
Exilium . . .	C.	118	1	5	
Expelled . . .	G.	215	8	3	
Expert . . .	G.	771	6	5	
Exploit . . .	G.	603	9	4	
Extinct . . .	M.	961	7	5	
Fabia . . .	F.	94	4	3	
Fabyan . . .	C.	143	8	2	
Face . . .	G.	150	6	A.	
Face Value . .	G.	277	7	A.	
Fad . . .	G.	85	4	A.	
Fadan . . .	C.	135	9	4	
Fair Alice . .	F.	391	4	4	Alice, 101 = 2.
Fair Atalanta . .	M.	1173	3	A.	Atalanta, 883 = 1.
Fair Duchess . .	M.	367	7	5	Duchess, 77 = 5.
Fairholm . . .	F.	341	8	2	
Fair Jean . . .	F.	353	2	2	Jean, 63 = 9.
Fairland . . .	M.	375	6	A.	
Fair Meddler . .	F.	574	7	3	Meddler, 284 = 5.

Name.	Sex.	Value.	Remainder.	Age (figures indicate years and "A" aged).	Remarks.
Fair Nell . . .	M.	380	2	5	Nell, 90=9.
Fair Penitent . .	F.	1190	2	4	Penitent, 990=9.
Fair Persian . .	M.	920	2	6	Persian, 630=9.
Fair Rebel . .	F.	542	2	3	Rebel, 252=9.
Fair Start . .	C.	1351	1	4	Start, 1061=8.
Fairy Field . .	G.	242	⅓	6	Fairy, 300=3.
Fairy Lamp . .	M.	451	1	5	Lamp, 151=7.
Fairy Queen II. .	M.	380	2	A.	Queen, 80=8.
Fairy Saint . .	M.	820	1	5	Saint, 520=7.
Faisan d'Or . .	C.	353	2	4	
Faith Cure . .	G.	731	2	6	Cure, 236=2.
Falconbridge . .	C.	384	8	3	Bridge, 205=7.
False Report . .	M.	1067	5	6	Report, 896=5.
Fame II. . . .	M.	130	4	5	
Fame and Fortune .	F.	520	7	2	Fame, 130=4.
Fanciful . . .	M.	301	4	A.	
Fancy . . .	M.	201	3	A.	
Fancy Boy . .	G.	215	8	5	Boy, 14=5.
Fancy Shot . .	G.	903	3	5	Shot, 702=9.
Fanny II. . .	M.	141	6	5	
Fantastic . . .	G.	1012	4	3	
Farina . . .	F.	332	8	2	
Farndale (By) . .	G.	375	6	3	
Farndon Ferry . .	G.	685	1	4	Ferry, 300=3.
Farringford . .	G.	635	5	3	
Fascinator . .	G.	792	9	A.	
Fascinate . . .	F.	601	7	3	
Fast Castle . .	C.	652	4	3	Castle, 111=3.
Faster . . .	G.	741	3	A.	
Fasting Man . .	G.	702	9	A.	Man, 91=1.
Fatalist . . .	C.	972	9	4	
Father Wolf . .	C.	401	5	4	Wolf, 116=8.
Faugh a Ballagh II.	G.	116	8	4	
Favoloo (By) . .	F.	203	5	3	
Fawler . . .	G.	316	1	A.	.
Feather . . .	G.	294	6	5	
Fedamore . .	G.	341	8	A.	
Felise . . .	F.	190	1	4	
Felstead . . .	F.	594	9	3	
Feltrim . . .	C.	760	4	5	
Fencote . . .	H.	566	8	A.	
Fenris . . .	G.	400	4	3	
Fer	C.	290	2	3	
Fergus Belle . .	M.	402	6	6	Bell, 42=6.

Name.	Sex.	Value.	Remainder.	Age (figures indicate years and "A" aged).	Remarks.
Fergus Foam . .	M.	486	9	5	Foam, 126=9.
Fermayne . .	M.	380	2	5	
Fermoyle . .	C.	352	1	3	
Fernandez (By) .	C.	402	6	3	
Ferndale II. . .	G.	374	5	4	
Fernia . . .	M.	341	8	5	
Ferriera . . .	M.	501	6	A.	
Ferrule . . .	F.	326	2	4	
Ferryboat . .	C.	708	6	3	
Ferry Lass . .	F.	391	4	4	Lass, 91=1.
Festivity . . .	G.	1040	5	A.	
Festivity II. . .	M.	1040	5	6	
Feu de Paille . .	C.	225	9	3	
Feu Follet . .	F.	213	6	4	
Ffarwel . . .	M.	327	3	5	
Fiddler's Green .	G.	654	6	A.	
Fiddlestring . .	G.	844	7	A.	
Fidra . . .	F.	285	6	3	
Fifeshire II. . .	G.	660	3	6	
Fighting Furley .	H.	870	6	5	Furley, 320=5.
Filassier . . .	G.	381	3	A.	
Filliford . . .	H.	394	7	A.	
Fillipeen . . .	C.	250	7	4	
Fine Weather . .	C.	350	8	2	
Finish . . .	H.	430	7	A.	
Finnebrogue . .	M.	368	8	5	
Finvara . . .	C.	412	7	3	
Fiorino . . .	G.	352	1	A.	
Firebrand II. . .	G.	547	7	A.	
Fire Fay . . .	C.	380	2	4	
Firefiend . . .	G.	434	2	A.	
Fireguard . .	H.	515	2	5	
Fire Island . .	F.	386	8	4	
Fireman . . .	C.	380	2	3	
First Attempt . .	C.	1391	5	4	Attempt, 851=5.
First Avenue . .	G.	687	3	4	Avenue, 147=3.
First Off . . .	F.	622	1	4	Off, 82=1.
First Principal .	H.	1040	5	5	Principal, 500=5.
Fishguard . .	G.	606	3	A.	
Fistiana . . .	M.	602	8	5	
Fits and Starts .	C.	1716	6	4	
Fitz George . .	G.	695	2	A.	
Fitzgerald . .	G.	734	5	A.	
Fitz Monarque .	C.	804	3	4	

Name.	Sex.	Value.	Remainder.	Age (figures indicate years and "A" aged).	Remarks.
Fitz Simon (By) .	G.	637	7	4	
Fitz Stuart .	C.	1554	6	3	
Fiume . .	H.	136	1	5	
Five of Hearts .	M.	907	7	A.	Five, 160=7.
Five Scents .	G.	740	2	5	
Flag . . .	C.	131	5	2	
Flageolet II. .	C.	587	2	3	
Flamenco .	H.	237	3	6	
Flavus . .	H.	251	8	A.	
Flaw . .	H.	112	4	5	
Flaxen Princess .	F.	821	2	4	Princess, 460=1.
Fleche . .	F.	420	6	3	
Fleeting Love .	F.	700	7	4	Love, 110=2.
Fleur d'Ete .	F.	744	6	4	Fleur, 320=5.
Flexible .	G.	232	7	4	
Flinton .	G.	610	7	A.	
Flipper II. .	G.	312	6	A.	
Flirtilla II. .	M.	741	3	6	
Floors . .	F.	376	7	2	
Floradora II. .	M.	528	6	A.	
Floreal .	M.	356	5	5	
Florence Reddy .	M.	660	3	5	Florence, 436=4.
Florentine (By) .	F.	836	8	3	
Florescene .	F.	446	5	3	
Floriage .	C.	329	5	3	
Florianus .	G.	437	5	3	
Florid .	C.	320	5	2	
Floriform .	C.	638	8	4	
Florimel .	M.	396	9	A.	
Florin .	C.	366	6	3	
Florinda .	F.	371	2	2	
Florist .	H.	776	2	A.	
Florodora .	F.	533	2	4	
Flourish .	G.	616	4	5	
Flower .	F.	318	3	2	
Flower of Spring .	C.	809	8	3	Flower, 318=3.
Flower of York .	M.	631	1	A.	Flower, 318=3.
Flutterer .	G.	910	1	5	
Flying Amphion .	C.	371	2	3	Amphion, 191=2.
Flying Crow .	C.	406	1	4	Crow, 226=1.
Flying Deer .	C.	394	7	4	Deer, 214=7.
Flying Fad .	G.	265	4	6	Fad, 85=4.
Flying Fancy II. .	M.	321	6	A.	Fancy, 141=6.
Flying Hackle .	H.	236	2	5	Hackle, 56=2.

Name.	Sex.	Value.	Remainder.	Age (figures indicate years and "A" aged).	Remarks.
Flying Hampton .	H.	676	1	A.	Hampton, 496=1.
Flying Jib . .	F.	185	5	3	Jib, 5.
Flying Lemur .	C.	460	1	3	Lemur, 280=1.
Flying Mist .	G.	680	5	6	Mist, 500=5.
Flying Peggy .	M.	300	3	5	Peggy, 120=3.
Flying Shot .	G.	882	9	6	Shot, 702=9.
Fokien . . .	G.	166	4	6	
Folkland . .	G.	191	2	4	
Folklore . .	F.	342	9	3	
Folkmote . .	C.	552	3	3	
Fonda . .	F.	137	2	2	
Fool . . .	H.	116	8	5	
Footlight . .	G.	910	1	6	
Footpath . .	C.	966	3	3	
Forcemeat .	G.	796	4	6	
Foreman . .	G.	376	7	3	
Forest Deer .	G.	966	3	4	Deer, 214=7.
Forest Lass .	M.	843	6	A.	Lass, 91=1.
Forfarshire .	H.	1062	9	5	
Forgetful . .	F.	820	1	4	
Form . .	G.	322	7	A.	
Formby . .	F.	334	1	2	
Forse . .	H.	346	4	6	
For Shame .	G.	632	2	A.	
Fortification .	H.	1133	8	6	
Fortune Hunter .	G.	990	9	6	Hunter, 655=7.
Forty Winks .	F.	828	9	4	Winks, 136=1.
Fortunio (By) .	C.	754	7	2	
—— . .	F.	754	7	2	
Forward . .	H.	492	6	5	
Fosco . .	G.	168	6	A.	
Fossicker .	G.	362	2	A.	
Fossil . .	M.	172	1	6	
Four . .	G.	286	7	A.	
Fourmi . .	F.	336	3	4	
Fowling-piece .	C.	336	3	3	Piece, 150=6.
Fox-bane . .	F.	254	2	2	Bane, 62=8.
Foxcatcher .	C.	416	2	4	Catcher, 224=8.
Foxhunter .	C.	847	1	4	Hunter, 655=7.
Francisque .	C.	471	3	3	
Frank Ash .	G.	662	5	3	Ash, 311=5.
Frank Buckle .	C.	403	7	4	Buckle, 52=7.
Franks . .	G.	411	6	4	
Fred Stone .	H.	810	9	5	Fred, 294=6.

Name.	Sex.	Value.	Remainder.	Age (figures indicate years and "A." aged).	Remarks.
Free Bird	G.	496	1	A.	Bird, 206 = 8.
Freebooter II.	G.	898	7	5	
Free Companion	H.	541	1	5	Companion, 231 = 6.
Freedom	G.	334	1	6	
Free Fight	G.	770	5	A.	Fight, 480 = 3.
Freeland	G.	375	6	5	
Free Love	G.	400	4	5	Love, 110 = 2.
Freemason (By)	F.	441	9	2	
——	F.	441	9	2	
——	F.	441	9	2	
——	F.	441	9	2	
——	C.	441	9	2	
——	F.	441	9	2	
French Maid	F.	397	1	4	Maid, 54 = 9.
Frenchpark	H.	644	5	6	Park, 301 = 4.
Fresh Air	F.	801	9	3	Air, 211 = 4.
Freville	C.	400	4	4	
Friar Buck	G.	503	8	5	Buck, 22 = 4.
Friar John	G.	536	5	6	John, 55 = 1.
Friarlike	H.	531	9	A.	
Friar of the East	C.	1047	3	4	Friar, 481 = 4.
Friar Balsam (By)	C.	614	2	3	Friar, 481 = 4; Balsam, 133 = 7.
——	G.	614	2	A.	Friar, 481 = 4; Balsam, 133 = 7.
Friar Wash	C.	788	5	4	Friar, 481 = 4; Wash, 307 = 1.
Friar Tuck	C.	901	1	3	Friar, 481 = 4; Tuck, 420 = 6.
Friary	H.	491	5	A.	
Friday	G.	294	6	A.	
Friday II.	C.	294	6	3	
Friendship	G.	724	4	6	
Frieze	G.	297	9	4	
Frivolity	M.	806	5	A.	
Frosty	H.	752	5	5	
Frozen Leagues	F.	463	4	4	Leagues, 120 = 3.
Frozen Out	G.	750	3	4	
Full Flavour	G.	501	6	6	Flavour, 391 = 4.
Full Gallop	H.	241	7	A.	Gallop, 131 = 5.
Full Hand	M.	170	8	5	Hand, 60 = 6.
Full of Luck	G.	241	7	5	
Full Ripe	M.	390	3	6	Ripe, 280 = 1.
Fusee II.	G.	113	5	A.	

Name.	Sex.	Value.	Remainder.	Age (figures indicate years and "A" aged).	Remarks.
Futen . . .	C.	546	6	2	
Fuzzy Wuzzy . .	G.	120	3	5	
Gabrielle d'Estrees .	F.	947	2	3	Gabrielle, 263 = 2.
Gadwall . . .	C.	63	9	4	
Gaffer Green .	G.	581	5	A.	Gaffer, 301 = 4.
Gag	F.	41	5	4	
Gailes . . .	C.	120	3	4	
Gainsborough . .	G.	350	8	5	
Gairloch . . .	M.	282	3	5	
Galashiels . .	H.	452	2	A.	
Galion . . .	C.	111	3	3	
Galivant . . .	G.	582	6	6	
Gallant . . .	H.	501	6	6	
Gallant Child .	C.	538	7	3	Child, 37 = 1.
Gallant Lass .	F.	592	7	3	Lass, 91 = 1.
Gallatin . . .	H.	502	7	A.	
Gallia . . .	M.	62	8	6	
Gallinule . . .	C.	147	3	2	
——— . . .	C.	147	3	3	
Gallon . . .	G.	101	2	6	
Galloping Girdle .	C.	455	5	2	Girdle, 254 = 2.
Galloping Helen .	F.	296	8	3	Helen, 95 = 5.
Galloping Jack .	C.	225	9	2	Jack, 24 = 6.
Galloping Lad (By) .	C.	236	2	2	Lad, 35 = 8.
——— . . .	C.	236	2	2	Lad, 35 = 8.
——— . . .	C.	236	2	2	Lad, 35 = 8.
——— . . .	G.	236	2	3	Lad, 35 = 8.
Gallops and Stays .	H.	723	3	2	Gallops, 191 = 2 ; Stays, 477 = 9.
Gallotte . . .	F.	453	3	3	
Gallowglass .	G.	168	6	5	
Galveston .	H.	651	3	5	
Galway Lights . .	F.	557	8	2	
Gambler . . .	G.	293	5	A.	
Game Bird .	C.	276	6	2	Bird, 206 = 8.
Game Chick .	F.	93	3	3	Chick, 23 = 5.
Gamecock II. .	G.	112	4	A.	Cock, 42 = 6.
Gam Hen II. .	M.	135	9	A.	Hen, 65 = 2.
Gangbridge .	G.	296	8	A.	
Ganymede .	G.	135	9	3	
Garb Or . . .	C.	425	2	4	
Gardaloo . . .	G.	262	1	5	
Gardenhurst .	G.	940	4	4	

Name.	Sex.	Value.	Remainder.	Age (figures indicate years and "A" aged).	Remarks.
Gardenia . . .	F.	296	8	4	
Garendon. . .	G.	335	2	5	
Garnish II. . .	G.	571	4	5	
Garrison Belle . .	F.	320	5	2	Belle, 42=6.
Garryowen . .	C.	298	1	3	Garry, 231=6.
Garsdale . . .	G.	325	1	4	
Garter Knight . .	G.	1271	2	4	Knight, 450=9.
Garterless . .	M.	921	3	6	
Garvaghy. . .	C.	332	8	2	
Gascony . . .	C.	161	8	3	
Gasparilla . .	M.	393	6	5	
Gatacre . . .	G.	642	3	2	
Gatherer . . .	C.	425	2	3	
Gauntlet . . .	G.	912	3	A.	
Gay	F.	30	3	3	
Gay and Sweet .	G.	591	6	A.	Gay, 30=3.
Gay Consul . .	H.	192	3	5	Consul, 162=9.
Gay Dog . . .	C.	56	2	3	Dog, 26=8.
Gayfield Lass . .	F.	245	2	4	Lass, 91=1.
Gay Lad . . .	C.	65	2	4	Lad, 35=8.
Gay Lad . . .	G.	65	2	4	Lad, 35=8.
Gaylock II. . .	G.	82	1	A.	
Gaylord . . .	C.	266	5	4	Lord, 236=2.
Gay Lord Quex .	G.	376	7	A.	Quex, 110=2.
Gay Reveller (By) .	F.	560	2	4	Reveller, 530=8.
Gean Tree . .	C.	763	7	4	Jean, 63=9.
Geelong . . .	G.	132	6	4	
Gelert . . .	G.	660	3	6	
Gell Pool . . .	G.	176	5	3	Gell, 60=6.
Gem . . .		53	8	A.	
General . . .	C.	283	4	4	
General II. . .	G.	283	4	4	
General III. . .	G.	283	4	A.	
General Buller .		515	2	A.	Buller, 232=7.
General Cronje .	C.	568	1	2	Cronje, 285=6.
Genius . . .	G.	123	6	4	
Genoa . . .	C.	140	5	2	
Gentiana . . .	F.	515	2	4	
Gentilhomme . .	C.	540	9	3	
Gentle Hilda . .	M.	523	1	6	Hilda, 40=4.
Gentlemen Joe .	G.	582	6	3	
George . . .	G.	208	1	6	
George Fordham .	G.	543	3	5	George, 208=1.
Georgina . . .	F.	270	9	4	

Name.	Sex.	Value.	Remainder.	Age (figures indicate years and "A" aged).	Remarks.
Geraldine II. . .	M.	298	2	A.	
Gerb . . .	C.	205	7	3	
Germaine. . .	M.	303	6	A.	
Germanicus .	G.	374	5	5	
Gertrude . .	M.	830	2	6	
Gervas (By) .	G.	344	2	2	
——— . .	F.	344	2	2	
——— . .	G.	344	2	2	
——— . .	G.	344	2	3	
Ghost Dance .	C.	511	7	2	Dance, 115=7.
Giantess . .	M.	534	3	A.	
Giddy Goat .	G.	460	1	4	Goat, 426=3.
Giganteum (By)	G.	524	2	3	
——— . .	C.	524	2	2	
Giglio . .	H.	59	5	6	
Gilbert . .	H.	652	4	6	
Gilbert Handley	C.	752	5	2	Gilbert, 662=5.
Gillette . .	F.	460	1	4	
Gilfach . .	G.	151	7	4	
Gillette . .	C.	640	1	3	
Gillie II. . .	G.	60	6	A.	
Gingerbread .	F.	472	4	2	
Ginger Jack .	G.	280	1	A.	Jack, 24=6.
Gingre . .	G.	273	3	6	
Gipsy Girl .	M.	403	7	6	Girl, 250=7.
Girasol . .	C.	317	2	3	
Girlie II. . .	M.	260	8	6	
Gironde . .	F.	286	7	3	
Girsha . .	F.	521	8	2	
Girton Girl .	F.	920	2	2	Girl, 250.
Girton M.A. .	F.	732	3	4	Girton, 670=4.
Glacier . .	C.	321	6	3	
Gladness . .	M.	175	4	A.	
Gladwyn . .	G.	111	3	3	
Glamorgan .	G.	363	3	A.	
Glasalt . .	F.	542	2	4	
Glasshampton .	G.	607	4	2	
——— . .	C.	607	4	2	
——— . .	F.	607	4	4	
——— . .	F.	607	4	3	
——— . .	M.	907	4	5	
——— . .	F.	607	4	3	
Glass Jug. .	F.	134	7	3	Jug, 23=5.
Gleeful . .	F.	170	8	4	

Name.	Sex.	Value.	Remainder.	Age (figures indicate years and "A" aged).	Remarks.
Glena Bay . . .	M.	123	6	6	Bay, 12 = 3.
Glenairlie . . .	G.	360	9	3	
Glenap . . .	C.	113	5	4	
Glenart . . .	M.	711	9	6	
Glenathol . . .	F.	546	6	2	
Glen Choran . .	H.	387	9	6	Glen, 110 = 2.
Glen Dixon . .	F.	244	1	4	Glen, 110 = 2.
Glen Doll . .	C.	146	2	3	Glen, 110 = 2.
Glengariffe . .	G.	411	6	A.	
Glengyle . . .	C.	170	8	4	
Glenile . . .	H.	150	6	5	
Glenluce . . .	C.	210	3	4	
Glenmalur . .	H.	391	4	5	
Glenrocky . .	G.	342	9	6	
Glenrose . . .	G.	323	8	5	
Glen Royal . .	G.	352	1	A.	Glen, 110 = 2.
Glen Royston . .	C.	832	4	3	Glen, 110 = 2.
Glentworth . .	G.	1121	5	A.	
Glenvannon (By) .	G.	291	3	6	
Glenville . . .	M.	230	5	A.	
Glenwood (By) . .	G.	120	3	2	
——— . . .	F.	120	3	2	
Glimpse . . .	G.	150	6	3	
Glitters II. . .	G.	710	8	6	
Glorious Uncertainty	G.	1497	3	3	Uncertainty, 1171 = 1.
Glory Hole . .	G.	307	1	6	Glory, 266 = 5.
Gobo . . .	G.	34	7	A.	
Godchild . . .	F.	63	9	2	
Go Easy . . .	M.	54	9	6	
Gogo . . .	F.	52	7	4	
Goldayr . . .	C.	270	9	3	
Gol Dee . . .	F.	70	7	4	
Golden Age . .	H.	124	7	A.	Age, 14 = 5.
Golden Apple . .	M.	153	9	5	Apple, 43 = 7.
Golden Bay . .	H.	122	5	5	Bay, 12 = 3.
Golden Cabin . .	F.	183	3	3	Cabin, 73 = 1.
Golden Crown (By) .	C.	386	8	2	Crown, 276 = 6.
——— . . .	G.	386	8	2	Crown, 276 = 6.
Golden Dollar . .	M.	346	4	6	Dollar, 236 = 2.
Golden Eye . .	C.	121	4	2	Eye, 11 = 2.
Golden Fruit . .	F.	796	4	4	Fruit, 686 = 2.
Golden Furrow . .	H.	396	9	6	Furrow, 286 = 7.
Golden Goblet . .	C.	556	7	3	Goblet, 464 = 5.
Golden Hope II. .	G.	201	3	A.	Hope, 91 = 1.

Name.	Sex.	Value.	Remainder.	Age (figures indicate years and "A" aged).	Remarks.
Golden Knot	F.	562	4	4	Knot, 452 = 2.
Golden Lute	F.	546	6	3	Lute, 436 = 4.
Golden Owl	C.	147	3	2	Owl, 37 = 1.
Golden Penny	F.	260	8	2	Penny, 150 = 6.
Golden Queen	F.	190	1	3	Queen, 80 = 8.
Golden Ridge	G.	313	7	A.	Ridge, 203 = 5.
Golden Rule	G.	346	4	A.	Rule, 236 = 2.
Golden Sally	M.	211	4	5	Sally, 101 = 2.
Golden Song	H.	242	8	A.	Song, 132 = 6.
Golden Wand	F.	171	9	4	Wand, 61 = 7.
Golden Wedding	C.	200	2	4	Wedding, 90 = 9.
Golden Wishes	H.	476	8	5	Wishes, 366 = 6.
Golden Wren	F.	370	1	4	Rain, 260 = 8.
Goldfinch II.	G.	193	4	6	
Goldfinder II.	G.	394	7	6	Finder, 334 = 1.
Gold Florin	C.	426	3	4	Florin, 366 = 6.
Goldloan	G.	146	2	4	
Gold Lock	C.	112	4	2	Lock, 52 = 7.
Gold Mark	F.	321	6	4	Mark, 261 = 9.
Goldrush	F.	560	2	3	Rush, 500 = 5.
Goldwasher	G.	567	9	A.	
Gollanfield	G.	226	1	6	
Gondola	F.	113	5	2	
Gone Coon (By)	F.	148	4	2	
Gonzalez	C.	127	1	4	
Good Girl II.	M.	274	4	5	Girl, 250 = 7.
Good Luck	G.	74	2	6	Luck, 50 = 5.
Good Match	C.	68	5	3	Match, 44 = 8.
Good Morning	C.	386	8	4	Morning, 362 = 2.
Goodrest	C.	694	1	4	
Goodshot	G.	726	6	A.	Shot, 702 = 9.
Good Tidings	F.	558	9	2	Tidings, 534 = 3.
Good-time	M.	464	5	5	Time, 440 = 8.
Good Trick	M.	644	5	5	Trick, 620 = 8.
Goodwill	H.	60	6	5	Will, 36 = 9.
Goody Two Shoes	F.	753	6	3	Shoes, 313 = 7.
Gooseberry	G.	308	2	A.	
Goosey Gander	F.	371	2	4	Goosey, 96 = 6.
Gordianus	C.	351	9	3	
Gorgonzola	G.	340	7	5	
Gosling	M.	182	2	A.	
Gossima	M.	123	6	5	
Gossip	M.	162	7	A.	
Gossoon II.	G.	138	3	A.	

Name.	Sex.	Value.	Remainder.	Age (figures indicate years and "A" aged).	Remarks.
Goudie . . .	G.	40	4	A.	
Gourgaud . .	C.	252	9	2	
Grace Swift . .	F.	866	2	4	Grace, 290=2.
Grace Trenton . .	F.	1400	5	2	Grace, 290=2.
Grafter . . .	G.	901	1	A.	
Grammont (By) .	G.	712	1	4	
Grand Attack . .	G.	697	4	A.	Attack, 422=8.
Grandborough . .	C.	485	8	3	
Grandchild . .	F.	312	6	3	
Granddaughter .	F.	881	8	3	
Grand Deacon . .	C.	359	8	4	
Grand Duke (By) .	F.	315	9	2	Duke, 40=4.
—— . . .	C.	315	9	2	Duke, 40=4.
Grand Idea . .	F.	301	4	2	Idea, 26=8.
Grand Marnier .	F.	776	2	2	Marnier, 501=6.
Grand Slam . .	C.	406	1	3	Slam, 131=5.
Grange . . .	M.	274	4	A.	
Grangefield . .	G.	398	2	5	
Granger . . .	G.	474	6	5	
Gratification . .	C.	1072	1	4	
Gratin . . .	H.	671	5	A.	
Grattan . . .	G.	671	5	6	
Gratwicke . .	G.	653	5	5	
Grave and Gay .	F.	395	8	3	Grave, 310=4 ; Gay, 30=3.
Gravitation . .	F.	1052	8	3	
Great Match . .	F.	674	8	3	Match, 44=8.
Greatorex . . .	C.	926	8	2	
Great-time . .	F.	1070	8	4	Time, 440=8.
Grecian . . .	G.	350	8	A.	
Greek Lass . .	M.	341	8	6	Lass, 91=1.
Greek Slave . .	G.	430	7	A.	Slave, 180=9.
Green Cherry . .	F.	503	8	4	Cherry, 223=7.
Green Dragoon .	G.	561	3	A.	Dragoon, 281=2.
Greenfinch . .	C.	413	8	2	
Green Fly . .	C.	400	4	4	Fly, 120=3.
Greenhill . . .	G.	315	9	A.	
Green Isle . .	M.	319	4	5	Isle, 39=3.
Greenlawn (By) .	F.	362	2	2	
Green May . .	M.	330	6	5	May, 50=5.
Greenmount . .	G.	776	2	A.	
Green Road . .	C.	490	4	3	Road, 210=3.
Green Stone . .	C.	796	4	2	Stone, 516=3.
Green Witch . .	M.	289	1	5	Witch, 9=9.

Name.	Sex.	Value.	Remainder.	Age (figures indicate years and "A" aged).	Remarks.
Greta . . .	F.	631	1	4	
Gretna Green . .	F.	961	7	3	Gretna, 681 = 6.
Greybridge . .	C.	435	3	4	
Grey Friars (By) .	G.	771	6	2	Grey, 230 = 5.
—— . . .	F.	771	6	2	Grey, 230 = 5.
—— . . .	F.	771	6	3	Grey, 230 = 5.
Grey Child . .	C.	267	6	4	Child, 37 = 1.
Grey Leaf . .	F.	350	8	2	Leaf, 120 = 3.
Greyling . . .	G.	330	6	5	
Grey Lord . .	G.	466	7	A.	Lord, 236 = 2.
Greystoke . .	F.	717	5	2	
Greystone II. . .	G.	746	8	A.	
Grey Tick . .	H.	650	2	6	Tick, 420 = 6.
Gribou . . .	H.	228	3	A.	
Grimpo . . .	G.	346	4	A.	
Grim's Dyke . .	F.	301	4	3	
Grimwig . . .	G.	286	7	A.	
Griper . . .	G.	500	5	5	
Grisi . . .	F.	290	2	4	
Grogan . . .		296	8	A.	
Groomsman . .	C.	417	3	2	
Grudon . . .	H.	274	4	A.	
Grutli . . .	C.	660	3	3	
Gualala . . .	F.	89	8	2	
Guardsman . .	G.	376	7	A.	
Guerrilla . .	G.	251	8	6	
Guienne . .	F.	80	8	3	
Gun Club . .	C.	122	5/7	2	Club, 52 = 7.
Gun Court . .	C.	699	3	2	
Gunfire . .	F.	360	9	2	
Gunmaker . .	C.	331	7	3	Maker, 261 = 9.
Gunner . . .	H.	270	9	5	
Gunner II. . .	G.	270	9	A.	
Gustave Dore . .	H.	771	6	6	Gustave, 561 − 3.
Gusty . . .	G.	490	4	A.	
Guttersnipe . .	G.	810	9	5	
Gwnfa . . .	F.	157	4	3	
Gyp	M.	83	2	A.	
Gypo . . .	G.	89	8	2	
Habton . . .	G.	458	8	A.	
Hackett . . .	H.	436	4	6	
Hackle . . .	M.	56	2	5	
Hackler's Pride .	F.	600	6	2	Pride, 284 = 5.

Name.	Sex.	Value.	Remainder.	Age (figures indicate years and "A" aged).	Remarks.
Haggy . . .	F.	36	9	3	
Hainesby Rouge .	M.	643	4	5	Rouge, 506＝2.
Hair Dresser . .	F.	689	5	3	Dresser, 474＝6.
Haka . . .	H.	27	9	5	
Hakone . . .	F.	82	1	4	
Half a Look . .	G.	137	2	A.	Look, 50＝5.
Half-Caste . .	G.	567	9	5	
Half Chance . .	G.	200	2	5	
Half Hoop . .	G.	177	6	6	
Halidom . . .	H.	80	8	A.	
Halival . . .	G.	147	3	2	
Hall Caine . .	H.	117	9	A.	Hall, 37＝1.
Hallucination . .	F.	497	2	2	
Halo . . .	F.	42	6	3	
Halutos . . .	G.	502	7	A.	
Hamath . . .	C.	451	1	3	
Hammerkop . .	F.	348	6	2	
Hampstead . .	C.	520	7	2	
Hampton Boy . .	G.	510	6	5	Boy, 14＝5.
Hampton Light .	G.	926	8	A.	Light, 430＝7.
Hampton Lock .	H.	548	8	6	Lock, 52＝7.
Hampton Vine .	H.	626	5	A.	Vine, 130＝4.
Hand Gallop . .	F.	191	2	4	Gallop, 131＝5.
Handicapper . .	C.	283	4	4	
Hand in Hand . .	G.	171	9	2	Hand, 60＝6.
Handjar . . .	M.	264	3	A.	
Handshake . .	C.	390	3	3	
Hands Off . .	M.	202	4	A.	
Handspike . .	F.	220	4	4.	
Handy Man . .	C.	161	8	3	
Handy Man II. .	G.	161	8	4	
Haphazard . .	G.	303	6	3	
Haply . . .	F.	126	9	4	
Happy Boy . .	G.	32	5	4	Boy, 14＝5.
Happy Days . .	H.	39	3	5	Days, 21＝3.
Happy Lad . .	G.	53	8	5	Lad, 35＝8.
Happy Land II. .	G.	103	4	4	Land, 86＝8.
Happy Man . .	H.	109	1	5	Man, 91＝9.
Happy Match . .	F.	62	8	3	Match, 44＝8.
Happy Medium .	C.	122	5	3	Medium, 104＝5.
Happy Slave . .	C.	198	9	3	Slave, 180＝9.
Harbour . . .	H.	408	3	A.	
Harbour Master .	G.	1109	2	3	Master, 701＝8.
Hard Cash . .	F.	531	9	3	Cash, 321＝6.

Name.	Sex.	Value.	Remainder.	Age (figures indicate years and "A." aged).	Remarks.
Hard Labour . .	F.	443	2	2	Labour, 233=8.
Hard Luck . .	F.	260	8	4	Luck, 50=5.
Hardwick II. . .	G.	230	5	A.	
Haresfield . .	G.	399	3	4	
Harkaway II. . .	G.	243	9	A.	
Harlequin . .	G.	316	1	6	
Harling . . .	G.	306	9	A.	
Harry Halliard . .	C.	466	7	3	Harry, 216=9.
Hartfield . . .	G.	790	7	2	
Harvest Home II. .	G.	807	6	5	Home, 51=6.
Harvesting . .	G.	826	7	A.	
Harvest Money .	H.	856	1	A.	Money, 100=1.
Haskeval . . .	H.	207	9	5	
Haste . . .	G.	475	7	A.	
Hastoe . . .	C.	472	4	4	
Hastie Queen .	F.	556	7	2	Queen, 80=8.
Hatchel . . .	G.	39	3	A.	
Hathor . . .	F.	611	8	3	
Haut en bas . .	F.	464	5	4	
Havant . . .	G.	536	5	5	
Havoc II. . .	G.	106	7	A.	
Hawkeye (By) . .	F.	38	2	3	
Hawstead . .	G.	481	4	5	
Hay Presto . .	F.	380	2	4	Hay, 15=6.
Hazelhuhn . .	M.	104	5	5	
Hazledene . .	F.	107	8	4	
Hazlemere (By) .	C.	293	5	3	
Hazlenut II. . .	G.	493	7	5	
Headpiece . .	G.	169	7	5	
Heartache . .	F.	637	7	4	
Heart of Gold . .	F.	747	9	4	Heart, 606=3.
Hearwood . .	H.	225	9	6	
Heathcote Belle .	F.	888	6	5	Belle, 42=6.
Heather Fire . .	F.	509	5	3	
Heathvoe . . .	M.	506	2	A.	
Heaven's Light .	G.	635	5	A.	Light, 430=7.
Heavy Opera . .	C.	388	1	3	Opera, 283=4.
Hebe . . .	F.	17	8	4	
Heckler . . .	G.	265	4	A.	
Hector II. . .	G.	635	5	A.	
Hedera . . .	M.	230	5	5	
Heir Male . .	H.	295	7	A.	
Helen Macgregor .	F.	585	9	4	Helen, 95=5.
Helen Margaret .	M.	966	3	A.	Helen, 95=5.

9

Name.	Sex.	Value.	Remainder.	Age (figures indicate years and "A" aged).	Remarks.
Helenopont . .	C.	643	4	3	
Helium . . .	G.	85	4	A.	
Hellifield . . .	C.	169	7	2	
Hellin . . .	C.	95	5	3	
Henderson . .	G.	379	1	6	
Hendra . . .	F.	270	9	2	
Hengler's Pride .	H.	659	2	5	Pride, 284=5.
Herculanum II. .	C.	346	4	3	
Hercules II. . .	G.	272	2	6	
Her She Is Punches- town	M.	117	2	6	
Her Ladyship II. .	M.	630	9	5	
Hermana . . .	M.	297	9	5	
Herminius . .	H.	355	4	A.	
Herodias . . .	F.	295	7	3	
Hesitation . .	F.	773	8	4	
Hesse . . .	F.	85	4	2	
Hestercombe . .	F.	741	3	4	
Hidden Love . .	G.	169	7	5	Love, 110=2.
Hidden Treasure .	F.	876	3	4	Treasure, 817=7.
Higgler . . .	G.	255	3	6	
High Glee . .	G.	75	3	A.	Glee, 60=6.
High Jinks . .	M.	545	5	A.	Jinks, 530=8.
Highlander III. .	G.	300	3	6	
Highland Plaid .	M.	224	8	6	Plaid, 124=7.
High Stand . .	G.	530	8	4	Stand, 515=2.
Highway Songster .	G.	823	4	A.	Songster, 792=9.
Highwood . .	G.	25	7	A.	
Hilary . . .	F.	246	3	4	
Hill Green . .	G.	315	9	A.	
Hillmorton . .	G.	727	7	A.	
Hill of Bree . .	G.	328	4	6	Hill, 35=8.
Hill Star . . .	G.	696	3	5	Star, 661=4.
Hipi . . .	F.	95	5	3	
Hippolyta . .	M.	444	3	A.	
Hippy . . .	F.	17	8	4	
Hipswell . . .	C.	191	2	3	
His Exe . . .		133	7	5	
His Excellency .	H.	273	3	6	
His Grace . .	G.	302	5	A.	Grace, 290=2.
His Lordship . .	C.	628	7	3	
Hoar Frost . .	G.	953	8	A.	Frost, 742=4.
Hobnob . . .	G.	63	9	A.	
Hockey . . .	G.	37	1	5	

Name.	Sex.	Value.	Remainder.	Age (figures indicate years and "A" aged).	Remarks.
Hocus . . .	M.	91	1	6	
Hodson . . .	C.	121	4	4	
Hodge . . .	G.	10	1	2	
Hogarth . . .	G.	637	7	4	
Hollyhurst . .	G.	712	1	6	
Holme Lacey .	C.	152	8	2	Home, 51 = 6.
Holmfirth . .	M.	736	7	A.	
Holmwood . .	G.	61	7	4	
Holstein . . .	C.	701	8	4	
Holy Baroness .	F.	380	2	4	Baroness, 329 = 5.
Holy Heroine .	M.	322	7	A.	Heroine, 271 = 1.
Honest Indian .	G.	637	7	5	Indian, 115 = 7.
Honestus . . .	C.	582	6	4	
Honeycomb .	C.	131	5	3	
Honeyfall . .	C.	176	5	5	
Honeymoon II. .	M.	161	8	A.	
Honeymoon III. .	M.	161	8	A.	
Honeysuckle .	F.	175	4	3	
Honorarium .	M.	503	8	5	
Hooks and Eyes	F.	158	5	2	Hooks, 85 = 4.
Hoopoo . . .	F.	97	7	4	
Hopbine (By) .	C.	139	4	2	
—— . . .	G.	139	4	2	
Hope of the East	G.	657	9	3	Hope, 91 = 1.
Hope Temple .	M.	651	3	5	Hope, 91 = 1.
Hopetoun . .	C.	547	7	3	
Hopflower . .	F.	305	8	3	
Hop Fly . . .	F.	207	9	4	Fly, 120 = 3.
Hopgoblin . .	H.	191	2	5	
Hop the Twig .	C.	136	1	3	
Hop-vine . . .	C.	217	1	4	
Horicon . . .	G.	283	4	A.	
Hornet . . .	M.	667	1	5	
Hornpool . .	H.	373	4	A.	
Horse Chestnut .	G.	790	7	A.	Horse, 267 = 6.
Horsegill . .	H.	317	2	A.	
Horseplay . .	G.	387	9	A.	
Horton . . .	C.	657	9	5	
Hospitaller . .	C.	778	4	3	
Host . . .	C.	471	3	2	
Hot Sauce . .	C.	529	7	4	Sauce, 122 = 5.
Hottentot . .	G.	1269	9	5	
Howcleugh . .	G.	77	5	A.	
Howth . . .	G.	416	2	4	

Name.	Sex.	Value.	Remainder.	Age (figures indicate years and "A" aged).	Remarks.
Hoy	C.	21	3	4	
Hoya . . .	C.	22	4	3	
Hoylake . . .	M.	77	5	6	
H.R.H. . . .	C.	230	5	4	
Hubble Bubble .	F.	71	8	4	
Hudibras . . .	H.	212	5	5	
Hugath Lath .	G.	867	3	A.	Hugath, 431=8.
Hugh the Heron .	G.	300	3	5	Hugh, 21=3.
Hulcot . . .	H.	457	7	5	
Humerus . . .	G.	321	6	6	
Hungarian . .	C.	337	4	3	
Hunting Box .	G.	639	9	A.	Box, 114=6.
Hunting Morn .	H.	817	7	A.	Morn, 292=4.
Huree Babu . .		226	1	6	Huree, 215=8.
Hurricane II. . .	H.	285	6	A.	
Hurry II. . . .	M.	215	8	A.	
Hurry On . .	G.	267	6	5	
Hurst Park . .	F.	966	3	4	Park, 301=4.
Hutton . . .	G.	455	5	A.	
Hydraulic . .	C.	261	9	3	
Hyphrantes . .	G.	806	5	A.	
Ian	C.	52	7	4	
Iceland Moss .	F.	201	3	3	Maid, 54=9.
Ice Maiden . .	F.	166	4	3	Maiden, 104=5.
Ice Pudding . .	F.	216	9	4	Pudding, 154=1.
Ichi Ban . . .	H.	67	4	5	
Icon	C.	73	1	3	
Icy Lass . . .	F.	163	1	3	
Idalus . . .	G.	96	6	A.	
Idle Girl . . .	F.	286	7	4	Girl, 250=7.
Ikona . . .	G.	78	6	4	
Ile Bardelle . .	F.	288	9	3	Bardelle, 247=4.
Illinois II. . .	C.	98	8	3	
Imbroglio . .	H.	295	7	A.	
Imperialist . .	F.	831	3	3	
Imperial Light Horse	G.	1068	6	3	Horse, 267=6.
Imperious . .	F.	401	5	2	
Impious . . .	G.	191	2	6	
Impromptu . .	C.	452	2	3	
Incandescence .		317	2	5	
Incantation . .	F.	873	9	3	
Incense . . .	H.	231	6	5	
Incredible . .	M.	317	2	A.	

Name.	Sex.	Value.	Remainder.	Age (figures indicate years and "A" aged).	Remarks.
Indemnity . .	H.	565	7	5	
Indian Corn . .	M.	387	9	5	Corn, 272＝2.
Indian Ink . .	F.	186	6	4	Ink, 71＝8.
Indian Prince . .	C.	505	1	3	Prince, 390＝3.
Indigo . . .	F.	81	9	2	
Induction . .	F.	425	2	3	
In Front . .	F.	781	7	2	
Ingebrigt . .	G.	683	8	A.	
Ingram (By) .	F.	311	5	2	
Ingratitude . .	F.	685	1	2	
Inishfree . .	G.	641	2	5	
Initiated . .	F.	766	1	2	
Injustice . .	F.	574	7	2	
Inlaid . . .	G.	95	5	2	
Innellan . .	H.	141	6	5	
Innocence . .	H.	237	3	6	
Inquisitor . .	G.	428	5	A.	
Insulator . .	H.	742	4	3	
Intaglio . .	C.	518	5	3	
Intimidater . .	G.	1096	7	A.	
Intolerance . .	H.	793	1	A.	
Intruder . .	C.	662	5	3	
Invasion . .	C.	198	9	4	
Invercanny . .	G.	412	7	5	
Inwoods . .	F.	121	4	2	
Ione . . .	M.	67	4	A.	
Ipswich . .	C.	150	6	3	
Ireland Yet II. (By)	M.	705	3	5	
Ireton . .	C.	661	4	3	
Irish Boy . .	G.	516	3	5	Boy, 14＝5.
Irish Chief .	G.	595	1	6	Chief, 93＝3.
Irish Frieze .	C.	799	7	3	Frieze, 297＝9.
Irish Guard .	G.	728	8	6	Guard, 226＝1.
Irish Lass .	F.	593	8	4	Lass, 91＝1.
Irish Light II. .	G.	932	5	6	Light, 430＝7.
Irish Linen .	H.	632	2	5	Linen, 130＝4.
Irish Maid .	F.	556	7	3	Maid, 54＝9.
Irish Paddy .	G.	597	3	6	Paddy, 95＝5.
Irish Thistle .	G.	997	7	A.	Thistle, 495＝9.
Irish Witch .	F.	511	7	2	Witch, 9＝9.
Ironclad II. .	G.	307	1	4	
Isaac II. . .	H.	29	2	5	
Isinglass (By) .	H.	170	8	5	
——— . .	C.	170	8	2	

Name.	Sex.	Value.	Remainder.	Age (figures indicate years and "A" aged).	Remarks.
Isinglass (By) . .	G.	170	8	3	
Isis II. . . .	M.	122	5	A.	
Isla de Cuba . .	M.	76	4	6	Isla, 33=6.
Island Beau . .	H.	94	4	A.	Beau, 8=8.
Isle of Avalon . .	H.	275	5	6	Isle, 32=5.
Isle of Man . .	G.	204	6	4	Isle, 32=5.
Isle of Wight . .	F.	519	6	4	Isle, 32=5.
Ismail II.. . .	G.	211	4	A.	
Isobar (By) . .	C.	270	9	3	
Isobaric . . .	G.	290	2	4	
Isocheimal . .	H.	168	6	6	
Isocrates . . .	G.	757	1	4	
Isolation . . .	F.	449	8	2	
Isopod . . .	G.	152	8	6	
Issa	M.	62	8	6	
Issuant . . .	M.	518	5	6	
Isthmus . . .	F.	566	8	4	
Ivanhoe . . .	G.	143	8	A.	
Iveagh . . .	G.	92	2	A.	
Jabot . . .	C.	12	3	3	
Jackanapes . .	G.	216	9	4	
Jack Gubbins . .	C.	156	3	3	
Jack Hinton . .	H.	529	7	6	
Jack Tar . . .	G.	625	4	A.	Jack, 24=6.
Jack's the Lad . .	H.	133	7	5	
Jacobus . . .	H.	92	2	A.	
Jacqueline . .	F.	104	5	2	
Jacques Valbach .	G.	158	5	4	
Jamaica . . .	F.	75	3	2	
Jam Jar . . .	H.	248	5	5	
Janissary (By) . .	F.	324	9	2	
—— . . .	C.	324	9	3	
Japan . . .	C.	135	9	2	
Jason . . .	C.	114	6	2	
Jean Bart . . .	G.	266	5	3	Jean, 63=9.
Jean Bart II. . .	C.	266	5	4	Jean, 63=9.
Jeffie . . .	G.	103	4	3	
Jeffries . . .	C.	310	4	3	
Jennico . . .	C.	89	8	2	
Jenny Agnes . .	M.	214	7	5	Jenny, 73=1.
Jenny Hampton .	F.	569	2	4	Jenny, 73=1.
Jessy Skillan . .		244	1	4	Jessy, 83=2.
Jester . . .	G.	673	7	A.	

Name.	Sex.	Value.	Remainder.	Age (figures indicate years and " A " aged).	Remarks.
Jettatura . . .	F.	1021	4	4	
Jewel . . .		49	4	A.	
Jim . . .	G.	43	7	A.	
Jim's Cherry . .	G.	326	2	6	Cherry, 223=7.
Joanie . . .	F.	69	6	3	
Joaquina . . .	F.	81	9	3	
Jocose . . .	M.	38	2	5	
Jocular . . .	F.	255	3	2	
Jocus . . .	G.	89	8	2	
Joe the Marine .	H.	323	8	5	Joe, 9=9.
Joe Ullman . .	G.	130	4	A.	Joe, 9=9.
John . . .	G.	55	1	A.	
John Danby . .	C.	122	5	3	John, 55=1.
John Dory . .	G.	275	5	3	John, 55=1.
John Peel . .	C.	175	4	2	John, 55=1.
Johnny Mack . .	G.	126	9	5	Johnny, 65=2.
Johnstown Lass .	M.	656	8	5	Lass, 91=1.
Jo I so I . .	G.	97	7	6	
Joker . . .	G.	229	4	A.	
Joker II. . .	G.	229	4	6	
Joker III. . .	G.	229	4	A.	
Jolly Joker . .	C.	274	4	4	Joker, 229=4.
Jolly Knight . .	G.	495	9	5	Knight, 450=9.
Jolly Roger II. .	H.	450	9	6	Roger, 405=9.
Jolly Sir George .	G.	513	9	A.	George, 208=1.
Jolly Tar . .	G.	646	7	6	Tar, 601=7.
Jonathan Wild . .	C.	555	6	2	Jonathan, 515=3.
Josephus . . .	G.	219	3	A.	
Joshua . . .	C.	312	6	3	
Jota . . .	F.	410	5	4	
Jouvence . . .	G.	209	2	5	
Jove . . .	G.	89	8	A.	
Jovial King . .	G.	219	3	4	
Jubert . . .	G.	605	2	5	
Jubilee Diamond .	H.	149	5	5	Diamond, 98=8.
Jubilee Jack . .	H.	75	3	5	Jack, 24=6.
Jubrai . . .	C.	216	9	3	
Jugfull . . .	G.	133	7	2	
Juggledale . .	C.	97	7	2	
Juggler (By) . .	F.	253	1	2	
—— . . .	F.	253	1	2	
—— . . .	C.	253	1	2	
Julia Franks . .	F.	461	2	2	
June Fly . . .	F.	179	8	4	Fly, 120=3.

Name.	Sex.	Value.	Remainder.	Age (figures indicate years and "A" aged).	Remarks.
Jungle Crow . .	C.	329	5	4	Crow, 226 = 1.
Juno . . .	M.	65	2	A.	
Jupiter II. . .	G.	689	5	A.	
Jurisprudence . .	C.	673	7	2	
Jusque au Bout .	C.	98	8	3	
Jusque La . .	F.	114	6	4	
Just Cause . .	C.	492	6	3	
Justice Royal . .	H.	765	9	6	Justice, 523 = 1.
Kadikoi . . .	C.	71	8	4	
Kaffa . . .	F.	102	3	4	
Kaffir Queen . .	F.	381	3	4	Queen, 80 = 8.
Kahouane . .	F.	92	2	4	
Kakimono . .	C.	143	8	3	
Kale	G.	60	6	A.	
Kalydor . . .	C.	267	6	3	
Kara . . .	C.	222	6	3	
Karakoul . . .	C.	278	8	3	
Karamanie . .	F.	323	8	3	
Karnak . . .	G.	292	4	6	
Karri . . .	M.	231	6	5	
Kate	M.	430	7	A.	
Kate Visto . .	F.	585	9	2	Kate, 430 = 7.
Katoomba . .	G.	470	2	4	
Katrine . . .	F.	681	6	4	
Kearsage . . .	C.	294	6	3	
Keen Blade . .	C.	126	9	4	Blade, 46 = 1.
Keendragh . .	F.	285	6	4	
Keepsake . . .	C.	210	3	3	
Kendal (By) . .	G.	114	6	3	
—— . . .	F.	114	6	2	
Kendal Brown . .	C.	372	3	3	Kendal, 114 = 6 ; Brown, 258 = 6.
Kendal Chief . .	C.	207	9	4	Chief, 93 = 3.
Kendal Glen . .	G.	224	8	A.	Glen, 110 = 2.
Kenloch . . .	C.	132	6	4	
Kenmure . . .	G.	336	3	A.	
Kennet . . .	F.	490	4	3	
Kennythorpe . .	C.	777	3	3	
Kenterdale . .	G.	724	4	6	
Kentish Glory . .	C.	1046	2	4	Glory, 266 = 5.
Kentshole . .	G.	581	5	A.	
Kenwyn . . .	G.	136	1	A.	
Kepi . . .	F.	120	3	2	

Name.	Sex.	Value.	Remainder.	Age (figures indicate years and "A" aged).	Remarks.
Kerlogue . . .	G.	276	6	A.	
Kerseymere . .	H.	540	9	5	
Keystone . . .	C.	546	6	4	
Key West . . .	G.	506	2	4	Key, 30=3.
Khaki II.. . .	M.	631	1	6	
Khasnadar . .	C.	915	6	3	
Khedive . . .	H.	620	8	A.	
Khiva Pass . .	C.	758	2	3	
Kibbor . . .	G.	222	6	5	
Kicksy Wicksy .	M.	206	8	5	Kicksy, 110=2.
Kilaroo . . .		257	5	A.	
Kilcheran. . .	C.	314	8	4	
Kilcooley . .	G.	116	8	6	
Kilderkin . .	M.	324	9	A.	
Kildona . . .	F.	111	3	2	
Kildrummie . .	G.	304	7	A.	
Kilfinnan . .	C.	231	6	3	
Kilgetty . . .	G.	490	4	4	
Kilglas . . .	C.	161	8	2	
Kilgrogan . .	G.	346	4	6	
Kilkerran. . .	H.	331	7	A.	
Killaidan . .	C.	114	6	3	
Killaloe . . .	G.	161	8	A.	
Killarkin . .	M.	321	6	5	
Killarney. . .	M.	311	5	5	
Killarue . . .	G.	257	5	5	
Killeen . . .	M.	110	2	5	
Killincarrick .	F.	341	8	3	
Killog . . .	F.	72	9	4	
Killucan . . .	C.	121	4	3	
Killure . . .	G.	256	4	6	
Killygally . .	F.	121	4	2	
Kilmallog . .	G.	142	7	5	
Kilmanners . .	C.	401	5	4	
Kilmantle . .	C.	571	4	3	
Kilmartin. . .	G.	741	3	A.	
Kilpatrick . .	G.	751	4	5	
Kilteel . . .	C.	490	4	3	
Kiltullagh . .	C.	481	4	5	
Kilwarlin (By) .	C.	337	4	3	
——— . . .	G.	337	4	2	
Kilworth . . .	G.	661	4	A.	
Kinbrace . .	C.	342	9	4	
Kineton Boy .	G.	544	4	4	Boy, 14=5.

Name.	Sex.	Value.	Remainder.	Age (figures indicate years and "A" aged).	Remarks.
Kingalic . . .	M.	141	6	5	
King Bonby . .	G.	156	3	A.	Bonby, 66=3.
King Cole . .		146	2	5	Cole, 56=2.
King David . .	G.	105	6	A.	David, 15=6.
King Eider . .	H.	305	8	6	Eider, 215=8.
Kingfield . . .	C.	214	7	4	
Kingfisher . .	H.	670	4	6	
Kinghampton .	H.	586	1	A.	Hampton, 496=1.
King of the Plains .	G.	415	1	6	King, 90=9.
King of the Severn .	H.	585	9	5	King, 90=9.
King Philip . .	G.	280	1	6	Philip, 190=1.
King Pippin . .	C.	222	6	3	Pippin, 132=6.
King Pluto . .	C.	612	9	2	Pluto, 522=9.
King Rover . .	G.	576	9	4	Rover, 486=9.
King's Birthday .	G.	718	7	2	Birthday, 621=9.
King's Bounty . .	C.	565	7	2	Bounty, 468=9.
Kingsbury Lad. .		397	1	A.	Lad, 35=8.
Kingscote . .	G.	576	9	4	
King's Courier .	H.	533	2	5	Courier, 436=4.
Kingsford . .	C.	434	2	4	
King's Head . .	G.	116	8	A.	Head, 19=1.
King's Idler . .	G.	333	9	5	Idler, 236=2.
King's Limner .	C.	417	3	3	
King's Lynn . .	H.	177	6	5	Lynn, 80=8.
King's Quest . .	H.	587	2	5	Quest, 490=4.
Kingston II. . .	C.	600	6	4	
Kinrara . . .	C.	472	4	3	
Kirby Hampton .	G.	728	8	6	Kirby, 232=7.
Kirkbride. . .	C.	446	5	2	
Kirkby Grange .	G.	526	4	4	Grange, 274=4.
Kirkham (By) . .	G.	285	6	3	
Kirkland . . .	G.	325	1	6	
Kirkmichael . .	C.	340	7	2	
Kirko . . .	G.	246	3	5	
Kirkthorpe . .	G.	927	9	3	
Kirkwall . .	G.	277	7	A.	
Kirtle Axe . .	M.	771	6	5	Axe, 121=4.
Kissing Cup . .	M.	250	7	A.	Cup, 100=1.
Kitchener II. . .	G.	273	3	4	
Kitsey Witsey . .	F.	966	3	4	Kitsey, 490=4.
Kittiwick . . .	F.	446	5	2	
Kitty . . .		430	7	4	
Kitty II. . .	M.	430	7	A.	
Kitty Asthore . .	F.	1102	4	3	Kitty, 430=7.

Name.	Sex.	Value.	Remainder.	Age (figures indicate years and "A" aged).	Remarks.
Kitty O'Brien . .	M.	699	6	5	Kitty, 430=7.
Kizil Kourgan . .	F.	354	3	3	
K. K. K. . .	G.	60	6	4	
Kladeradatch . .	G.	264	3	2	
Kleon . . .	H.	112	4	A.	
Klephte . . .	C.	550	1	3	
Klingsor . . .	C.	382	4	4	
Klotho . . .	F.	463	4	3	
Knicknack . .		141	6	3	
Knight . . .	G.	450	9	4	
Knightly . . .	C.	490	4	3	
Knight of Ruby (By)	F.	743	5	2	Knight, 450=9.
Knight of the Road .	C.	755	8	4	Knight, 450=9.
Knighton . . .	G.	500	5	5	
Knobstick . .	C.	534	3	4	
Knockbritt . .	G.	674	8	A.	
Knock Jon . .	G.	131	5	5	
Knocknagreana .	F.	404	8	2	
Knockshegowna .	C.	459	9	3	
Knocksouna . .	G.	189	9	6	
Koso-Pepsine . .	H.	382	4	6	
Kozak . . .	G.	44	9	A.	
Kroonstad . .	C.	741	3	2	
Kruger . . .	C.	446	5	3.	
Kruger III. . .	G.	446	5	A.	
Kuanos . . .	H.	137	2	A.	
Kumasi . . .	G.	131	5	A.	
Kummerbund . .	C.	316	1	3	
Kunstler . . .	C.	760	4	4	
Kurvenal . . .	G.	391	4	A.	
Kyle Rose . .	F.	273	3	3	
Kyoto . . .	G.	442	1	A.	
Kypie . . .	G.	120	3	A.	
Laarnce . . .	G.	342	9	5	
L'Abbe Morin (By) .	C.	339	6	2	Morin, 296=8.
La Bestia . . .	F.	514	1	4	Bestia, 483=6.
La Camargo . .	F.	319	4	4	Camargo, 288=9.
La Dragonne . .	F.	312	6	3	Dragonne, 281=2.
La Fille du Regiment	M.	465	6	A.	Fille, 12=3.
La Jonckee . .	F.	120	3	4	Jonckee, 89=8.
La Laide . . .	F.	75	3	2	Laide, 44=8.
La Loreley . .	F.	307	1	3	Loreley, 276=6.
La Lune . . .	M.	117	9	6	86=5.

Name.	Sex.	Value.	Remainder.	Age (figures indicate years and " A " aged).	Remarks.
La Moree . .	F.	287	8	3	Moree, 256 = 4.
La Napoule . .	F.	198	9	4	Napoule, 167 = 5.
La Noblesse . .	F.	189	9	3	Noblesse, 158 = 5.
La Plata . . .	M.	543	3	A.	Plata, 512 = 8.
La Poupee . .	M.	207	9	6	Poupee, 176 = 5.
La Prarie . .	C.	522	9	2	Prarie, 491 = 5.
La Quinta . .	F.	502	7	4	Quinta, 471 = 3.
La Uruguaya . .	M.	279	9	6	Uruguaya, 248 = 5.
La Valerie . .	F.	361	1	3	Valerie, 330 = 6.
La Vallierie . .	F.	351	9	3	Valliere, 320 = 5.
Lackford . . .	H.	335	2	A.	
Laconia . . .	F.	118	1	3	
Lactantius (By) .	F.	962	8	2	
——— . . .	F.	962	8	2	
——— . . .	F.	962	8	2	
Ladamos . . .	F.	142	7	2	
Ladas (By) . .	F.	96	6	3	
Ladies Field . .	F.	176	5	3	Field, 124 = 7.
Ladies' Man . .	C.	143	8	3	Man, 91 = 1.
Lad o' Wax . .	C.	128	2	3	Lad, 35 = 8.
Lady IV. . . .	M.	45	9	A.	
Lady Algy . .	F.	106	7	4	Algy, 61 = 7.
Lady Alice II. .	M.	146	2	A.	Alice, 101 = 2.
Lady Alicia . .	M.	147	3	5	Alicia, 102 = 3.
Lady Armie . .	M.	296	8	5	Armie, 251 = 8.
Lady Ashton . .	M.	806	5	5	Ashton, 761 = 5.
Lady Belhaven .	M.	223	7	5	Belhaven, 178 = 7.
Lady Bird III. .	M.	251	8	A.	Bird, 206 = 8.
Lady Blackwing .	F.	174	3	4	Blackwing, 129 = 3.
Lady Blair . .	F.	287	8	2	Blair, 242 = 8.
Lady Blayney . .	M.	147	3	A.	Blayney, 102 = 3.
Lady Blazes . .	M.	92	2	A.	Blazes, 47 = 2.
Lady Boreen . .	M.	313	7	5	Boreen, 268 = 7.
Lady Britta . .	F.	648	9	3	Britta, 603 = 9.
Lady Burgoyne .	F.	319	4	2	Burgoyne, 274 = 4.
Lady Car . . .	F.	266	5	2	Car, 221 = 5.
Lady Clifton II. .	M.	625	4	A.	Clifton, 580 = 4.
Lady Clyde . .	F.	109	1	3	Clyde, 64 = 1.
Lady Cole . .	F.	101	2	3	Cole, 56 = 2.
Lady Crafton . .	F.	796	4	2	Crafton, 751 = 4.
Lady Cull . .	F.	95	5	3	Cull, 50 = 5.
Lady Dern . .	F.	299	2	2	Dern, 254 = 2.
Lady Derry . .	M.	269	8	6	Derry, 224 = 8.
Lady Doneraile .	F.	355	4	2	Doneraile, 310 = 4.

Name.	Sex.	Value.	Remainder.	Age (figures indicate years and "A" aged).	Remarks.
Lady Drake	F.	279	9	2	Drake, 234=9.
Lady Eda	M.	61	7	6	Eda, 16=7.
Lady Egremont	F.	378	9	2	Egremont, 333=9.
Lady Elect	F.	516	3	4	Elect, 471=3.
Lady Fitz James	M.	636	6	A.	Fitz James, 501=6.
Lady Flanagan	M.	277	7	A.	Flanagan, 232=7.
Lady Flash	M.	456	6	A.	Flash, 411=6.
Lady Flight	M.	555	6	6	Flight, 510=6.
Lady Florence	F.	481	4	4	Florence, 436=4.
Lady Flush	M.	455	5	A.	Flush, 410=5.
Lady Fortune	M.	380	2	5	Fortune, 335=2.
Lady Gallinule	F.	182	2	3	Gallinule, 137=2.
Lady Glenwood	M.	165	3	A.	Glenwood, 120=3.
Lady Gordon	M.	321	6	6	Gordon, 276=6.
Lady Gracie	M.	336	3	5	Gracie, 291=3.
Lady Grand	F.	320	5	3	Grand, 275=5.
Lady Hackler	F.	301	4	3	Hackler, 256=4.
Lady Hartstown	M.	1167	6	5	Hartstown, 1122=6.
Lady Help	F.	170	8	2	Help, 125=8.
Lady Hermit	M.	690	6	A.	Hermit, 645=6.
Lady Hinton	M.	550	1	5	Hinton, 505=1.
Lady Hope	M.	136	1	5	Hope, 91=1.
Lady Hugo	M.	82	1	5	Hugo, 37=1.
Lady Iris	F.	316	1	4	Iris, 271=1.
Lady Isabel II.	M.	149	5	6	Isabel, 104=5.
Lady Janet	M.	509	5	5	Janet, 464=5.
Lady Killer	C.	295	7	4	Killer, 250=7.
Lady Lambkin	F.	186	6	4	Lambkin, 141=6.
Lady Linton II.	M.	575	8	A.	Linton, 530=8.
Lady Linthorpe	M.	812	2	A.	Linthorpe, 767=2.
Lady Lovely	F.	195	6	2	Lovely, 150=6.
Lady Lundy	M.	139	4	5	Lundy, 94=4.
Lady Macdonald	F.	195	6	3	Macdonald, 150=6.
Lady Macgregor	M.	535	4	5	Macgregor, 490=4.
Lady Malta	F.	517	4	3	Malta, 472=4.
Lady Marcion	F.	636	6	2	Marcion, 591=6.
Lady Marmiton	F.	776	2	2	Marmiton, 731=2.
Lady Massey	M.	156	3	5	Massey, 111=3.
Lady Maize	M.	102	3	6	Maize, 57=3.
Lady Mickleham	F.	190	1	2	Mickleham, 145=1.
Lady Min	M.	135	9	5	Min, 90=9.
Lady Nicotine	M.	581	5	5	Nicotine, 536=5.
Lady Novice	M.	237	3	6	Novice, 192=3.
Lady of Lyons	F.	276	6	4	Lyons, 150=6.

Name.	Sex.	Value.	Remainder.	Age (figures indicate years and " A " aged).	Remarks.
Lady of Milan . .	F.	256	4	3	Milan, 130 = 4.
Lady of the Lamp .	F.	291	3	2	Lamp, 151 = 7.
Lady Olive . .	M.	157	4	A.	Olive, 112 = 4.
Lady Ormac . .	F.	307	1	3	Ormac, 262 = 1.
Lady Paramount II.	F.	823	4	4	Paramount, 778 = 4.
Lady Patricia . .	F.	797	5	2	Patricia, 752 = 5.
Lady Penzance .	M.	303	6	5	Penzance, 258 = 6.
Lady Rivers . .	F.	585	9	4	Rivers, 540 = 9.
Lady's Realm . .	M.	332	8	5	Realm, 280 = 1.
Lady Royston . .	F.	767	2	4	Royston, 722 = 2.
Lady St. George .	F.	773	8	4	George, 208 = 1.
Lady Scattercash .	F.	1047	3	4	Scattercash, 1002 = 3.
Lady Sevington .	F.	715	4	2	Sevington, 670 = 4.
Lady Shamrock .	F.	608	5	4	Shamrock, 563 = 5.
Lady Stafford . .	M.	790	7	5	Stafford, 745 = 7.
Lady Steenie . .	F.	575	8	2	Steenie, 530 = 8.
Lady Sykes . .	M.	195	6	A.	Sykes, 150 = 6.
Lady Tout . .	M.	851	5	A.	Tout, 806 = 5.
Lady Valentine .	M.	666	9	5	Valentine, 621 = 9.
Lady Victor II. .	M.	745	7	A.	Victor, 700 = 7.
Lady White . .	F.	451	1	4	White, 406 = 1.
Lady Wilful . .	F.	191	2	4	Wilful, 146 = 2.
Lagado . . .	C.	62	8	3	
Lagoon . . .	G.	107	8	3	
Laguna . . .	F.	108	9	A.	
Laird . . .	G.	244	1	4	
Lakota . . .	M.	458	8	6	
Lambay . . .	M.	83	2	A.	
Lambel . . .	G.	113	5	A.	
Lambourne Belle .	F.	371	2	4	
Lammas . . .	G.	131	5	5	
Lampas . . .	G.	211	4	A.	
Lance . . .	G.	141	6	4	
Lancewood . .	C.	151	7	3	
Land we Live In .	C.	262	1	4	Land, 85 = 4.
Langholm . .	G.	152	8	4	
Lanoline . . .	F.	177	6	3	
Lantonius . .	C.	607	4	2	
Lapereau . . .	G.	327	3	3	
Laplander . .	G.	396	9	A.	
Lapsang . . .	G.	242	8	2	
Laracor . . .	G.	458	8	A.	
Larch Hill . .	G.	286	7	6	
Lariat . . .	M.	641	2	5	

Name.	Sex.	Value.	Remainder.	Age (figures indicate years and "A" aged).	Remarks.
Lass o' Glory . .	F.	364	4	3	Lass, 91 = 1.
Last Kiss . . .	F.	471	4	3	Kiss, 80 = 8.
Last Trick . .	G.	1111	4	4	Trick, 620 = 8.
Latitat . . .	F.	1232	8	2	
Laton . . .	M.	481	4	A.	
Laura . . .	F.	237	3	3	
Laura . . .	F.	237	3	4	
Laurel Vale . .	C.	392	5	2	Vale, 120 = 3.
Laurel Wreath . .	M.	887	5	6	Wreath, 615 = 3.
Lauriscope (By) .	F.	398	2	3	
Lauriston . .	G.	742	4	5	
Lavender Kid . .	G.	399	3	3	Kid, 24 = 6.
Lavengro . . .	C.	397	1	3	
Laveno (By) . .	F.	177	6	3	
—— . .	G.	177	6	2	
—— . .	F.	177	6	2	
Lawn . . .	F.	82	1	4	
Lawrence . . .	G.	352	1	5	
Lawrenny . .	G.	302	5	A.	
Le Bearnais II. .	C.	312	6	6	Bearnais, 272 = 2.
Le Blizon . . .	H.	131	5	6	Blizon, 91 = 1.
Le Buff . . .	H.	122	5	A.	Buff, 82 = 1.
Le Cadeau . .	M.	71	8	A.	Cadeau, 31 = 4.
Le Capucin . .	C.	251	8	3	Capucin, 211 = 4.
Le Chanoine . .	C.	531	9	3	Chanoine, 491 = 5.
Le Fere . . .	G.	330	6	A.	Fere, 290 = 2.
Le Firmament . .	C.	861	6	3	Firmament, 821 = 2.
Le Mandinet . .	C.	195	6	3	Mandinet, 155 = 2.
Le Mistral . .	C.	771	6	3	Mistral, 731 = 2.
Le Souvenir . .	C.	446	5	3	Souvenir, 406 = 1.
Le Var (By) . .	F.	321	6	2	Var, 281 = 2.
Le Vengeur . .	C.	410	5	4	Vengeur, 370 = 1.
Lea and Perrin .	C.	435	3	4	Lea, 40 = 4.
Leading Power .	M.	402	6	5	Power, 288 = 9.
Leatherstocking .	G.	796	4	A.	
Ledessan . . .	G.	165	3	A.	
Lee Metford . .	G.	774	9	6	Lee, 40 = 4.
Leewood . . .	G.	50	5	A.	
Legacy . . .	C.	130	4	4	
Leggan Hall . .	H.	147	3	A.	Hall, 37 = 1.
Leighton . . .	C.	490	4	2	
Leinon . . .	M.	140	5	6	
Leinster . . .	G.	750	3	4	
Leisure Hour . .	H.	454	4	A.	Hour, 207 = 9.

Name.	Sex.	Value.	Remainder.	Age (figures indicate years and "A" aged).	Remarks.
Lemco . . .	F.	106	7	3	
Lemoine . . .	H.	136	1	5	
Lena Dacre . .	M.	316	1	A.	Lena, 91 = 1.
Lena Rhodes . .	F.	361	1	4	Lena, 91 = 1.
Lencocyte . .	M.	586	1	A.	
Lendal . . .	G.	124	7	4	
Leominster (By) .	F.	796	4	2	
Leonid . . .	G.	100	1	6	
Leontodon . .	G.	556	7	A.	
Leo Tertius . .	C.	1006	7	4	Leo, 46 = 1.
Leslie . . .	C.	140	5	3	
Leslie Rose . .	F.	353	2	3	Leslie, 140 = 5.
Lesmarry . . .		351	9	A.	
Lesterlin (By) .	F.	780	6	2	
Let Me Go . .	G.	516	3	A.	
Lettre de Cachet .	G.	985	4	5	Lettre, 640 = 1.
Letzter Mohikaner .	C.	1369	1	3	Letzter, 447 = 6.
Levanter . . .	G.	771	6	A.	
Levens . . .	C.	230	5	4	
Levens Hall . .	C.	267	6	2	Hall, 37 = 1.
Leviathan . .	C.	586	1	4	
Levybub . . .	H.	134	8	A.	
Liberation . .	F.	583	7	3	
Liberte . . .	F.	632	2	4	
Liberty Bell . .	F.	684	9	4	Belle, 42 = 6.
Libya . . .	F.	43	7	3	
Lifebuoy . . .	G.	124	7	A.	
Light Blossom . .	M.	564	6	A.	Blossom, 134 = 8.
Light Hand . .	C.	490	4	4	Hand, 60 = 6.
Lights Out . .	F.	897	6	4	
Likely Bird . .	H.	296	8	5	Bird, 206 = 8.
Likewise . . .	F.	73	1	4	
Liliard . . .	F.	274	4	2	
Liliom . . .	C.	110	2	3	
Lily	G.	70	7	2	
Lily Castle . .	F.	181	1	2	Lily, 70 = 7.
Lily of the Valley .	F.	286	7	3	Lily, 70 = 7.
Lily Surefoot . .	M.	1056	3	5	Lily, 70 = 7.
Limerick II. . .	M.	290	2	A.	
Limone . . .	M.	126	9	6	
Limonum . . .	C.	166	4	2	
Limpsfield Lassie .	F.	435	3	3	Lassie, 91 = 1.
Linaria . . .	C.	292	4	3	
Linaro . . .	C.	287	8	3	

Name.	Sex.	Value.	Remainder.	Age (figures indicate years and " A " aged).	Remarks.
Lincoln Imp . . .	C.	271	1	2	Imp, 121 = 4.
Lindela . . .	M.	125	8	5	
Lindy . . .	F.	94	4	3	
Linkless . . .	G.	200	2	4	
Liquidator . . .	C.	655	7	3	
Lisette . . .	F.	500	5	3	
Lisinil . . .	F.	320	5	3	
Liskennett . .	G.	580	4	A.	
Lismany . . .	G.	191	2	A.	
Little Bear . .	G.	672	6	4	Bear, 212 = 5.
Little Ben . .	C.	522	9	4	Ben, 62 = 8.
Little Billee . .	G.	502	7	4	Billee, 42 = 6.
Little Bob . .	H.	466	7	A.	Bob, 6 = 6.
Little Brown Mouse	G.	824	5	5	Mouse, 106 = 7.
Little Champion .	H.	644	5	A.	Champion, 184 = 4.
Little Chapter .	F.	1144	1	3	Chapter, 684 = 9.
Little Chat . .	M.	864	9	5	Chat, 404 = 8.
Little Cicestrian II. .	G.	1240	7	A.	Cicestrian, 780 = 6.
Little Cis . .		580	4	5	Cis, 120 = 3.
Little Colonel .	G.	760	4	A.	Colonel, 300 = 3.
Little Connie .	F.	542	2	3	Connie, 64 = 1.
Little Dobbin .	C.	518	5	2	Dobbin, 58 = 4.
Little Dora .	M.	671	5	5	Dora, 211 = 4.
Little Dove .	M.	544	4	6	Dove, 84 = 3.
Little Eva .	M.	552	3	A.	Eva, 92 = 2.
Little Fidget .	F.	953	8	2	Fidget, 493 = 7.
Little Flower .	F.	778	4	2	Flower, 318 = 3.
Little Gert .	F.	1080	9	4	Gert, 620 = 8.
Little Gill .		493	7	5	Gill, 33 = 6.
Little Grey .	G.	690	6	3	Grey, 230 = 5.
Little Hercules .	G.	732	3	6	Hercules, 272 = 2.
Little Ida . .	F.	467	8	3	Ida, 7 = 7.
Little Jackdaw .	C.	490	4	2	Jackdaw, 30 = 3.
Little Jim II. .	G.	503	8	A.	Jim, 43 = 7.
Little Man .	G.	551	2	5	Man, 91 = 1.
Little Mary .	M.	711	9	5	Mary, 251 = 8.
Little May II. .	M.	510	6	6	May, 50 = 5.
Little Meg .	F.	530	8	2	Meg, 70 = 7.
Little Model .	G.	546	6	5	Model, 86 = 5.
Little Mot .	M.	902	2	A.	Model, 442 = 1.
Little Norah .	F.	717	6	4	Norah, 257 = 5.
Little Phyllis .	M.	630	9	A.	Phyllis, 170 = 8.
Little Queen .	F.	540	9	4	Queen, 80 = 8.
Little Sister II. .	M.	1180	1	A.	Sister, 720 = 9.

Name.	Sex.	Value.	Remainder.	Age (figures indicate years and "A" aged).	Remarks.
Little Star . .	M.	1121	5	A.	Star, 661 = 4.
Little Sweetheart .	M.	1572	6	5	Sweetheart, 1112 = 5.
Little Teddy .	G.	884	2	4	Teddy, 424 = 1.
Little Tim . .	C.	503	8	3	Tim, 43 = 7.
Little Willie .	C.	506	2	4	Willie, 46 = 1.
Lively . . .	F.	150	6	2	
Livermere .	M.	560	2	5	
Livorno . .	G.	368	8	A.	
Llama . . .	C.	73	1	2	
Llanmira . .	C.	323	8	4	
Llanthony (By) .	G.	547	7	2	
——— . .	F.	547	7	2	
——— . .	C.	547	7	2	
Loafer . .	G.	316	1	A.	
Loch Doon .	F.	112	4	4	
Lochinvar (By) .	M.	383	5	5	
Loch Leven .	C.	222	6	3	
Lockinge . .	H.	105	6	5	
Loddon . .	G.	84	3	A.	
Lofrasco . .	C.	403	7	2	
Logahoff . .	G.	148	4	A.	
Logan . .	G.	106	7	A.	
Lognes . .	C.	176	5	3	
Lohengrin .	G.	381	2	A.	
Lointaine . .	F.	542	2	3	
Lois II. . .	M.	46	1	A.	
Lollard . .	G.	266	5	A.	
Lomax . .	H.	157	4	A.	
London . .	G.	134	8	6	
London Assurance .	G.	751	4	3	Assurance, 617 = 5.
London Pride .	C.	418	4	4	Pride, 284 = 5.
Longbarrow .	G.	311	5	5	
Long Cecil .	G.	262	1	4	Cecil, 160 = 7.
Longnor . .	G.	352	1	6	
Longshoreman .	C.	698	5	4	
Longsword .	G.	372	3	6	Sword, 270 = 9.
Longthorpe .	C.	789	6	4	
Long Tom .	C.	544	4	3	Tom, 442 = 1.
Longy . .	H.	95	5	5	
Look Here .	F.	265	4	2	
Look Up . .	F.	131	5	3	
Loomgale . .	C.	136	1	3	
Looter . .	G.	636	6	3	
Lord Abbot .	C.	649	1	4	Abbot, 413 = 8.

Name.	Sex.	Value.	Remainder.	Age (figures in-dicate years and "A" aged).	Remarks.
Lord Ardington .	C.	961	7	4	Ardington, 725 = 5.
Lord Arravale . .	G.	558	9	A.	Arravale, 322 = 7.
Lord Audley . .	G.	282	3	A.	Audley, 46 = 1.
Lord Bob . . .	H.	242	8	A.	Bob, 6 = 6.
Lord Bobs . .	C.	302	5	4	Bobs, 66 = 3.
Lord Bruce . .	G.	504	9	A.	Bruce, 268 = 7.
Lord Burghley . .	C.	498	3	4	Burghley, 262 = 1.
Lord Carbine . .	C.	519	6	3	Carbine, 283 = 4.
Lord Danvers . .	G.	631	1	A.	Danvers, 395 = 8.
Lord Denman . .	G.	390	3	5	Denman, 154 = 1.
Lord Gardener .	G.	711	9	A.	Gardener, 475 = 7.
Lord Garvagh . .	H.	538	7	5	Garvagh, 302 = 5.
Lord Hope . .	G.	327	3	A.	Hope, 91 = 1.
Lord James . .	C.	340	7	4	James, 104 = 5.
Lord Kilkenny . .	H.	376	7	A.	Kilkenny, 140 = 5.
Lord Kyle . .	C.	296	8	2	Kyle, 60 = 6.
Lord Locksley . .	G.	388	1	6	Locksley, 152 = 8.
Lord Lorrha . .	G.	474	6	5	Lorrha, 238 = 4.
Lord Melton . .	C.	766	1	4	Melton, 530 = 8.
Lord Oriel . .	H.	483	6	5	Oriel, 247 = 4.
Lord President .	H.	1050	6	6	President, 814 = 4.
Lord Quex . .	C.	346	4	4	Quex, 110 = 2.
Lord Ronald II. .	G.	522	9	A.	Ronald, 286 = 7.
Lord Royston . .	C.	958	4	3	Royston, 722 = 2.
Lord Stavenger .	G.	1057	4	3	Stavenger, 821 = 2.
Lord Stavordale .	G.	1021	4	4	Stavordale, 785 = 2.
Lord Victor . .	C.	936	9	2	Victor, 700 = 7.
Lorenzaccio . .	H.	340	7	6	
Lorgnette . . .	F.	712	1	2	
Lorma . . .	F.	277	7	3	
Lorna Doon . .	F.	347	5	3	Lorna, 287 = 8.
Lony (By) . .	C.	242	8	2	
Lo'ruhamah . .	M.	293	5	6	
Lost Boy . . .	G.	506	2	3	Boy, 14 = 5.
Lost Chord . .	F.	718	6	2	Chord, 226 = 1.
Lottery . . .	G.	642	3	A.	
Loufogue . . .	C.	142	7	4	
Lough Allagh . .	M.	149	5	5	Lough, 112 = 4.
Lough-allo . .	C.	149	5	2	Lough, 112 = 4.
Lough Arrow . .	G.	329	5	A.	Lough, 112 = 4.
Loughran . .	G.	363	3	A.	
Louis . . .	G.	46	1	6	
Louis d'Or II. . .	G.	252	9	A.	Louis, 46 = 1.
Louisiana II. . .	F.	168	6	4	

Name.	Sex.	Value.	Remainder.	Age (figures indicate years and "A," aged).	Remarks.
Louisville . . .	G.	156	3	5	
Louis William . .	⎰C.	132	6	2	Louis, 46=1.
Loulou . . .	C.	72	9	5	
Loulou II. . .	C.	72	9	3	
Loupeau . . .	H.	122	5	5	
Louve . . .	F.	116	8	2	
Lovat . . .	C.	517	4	2	
Love Charm . .	C.	354	3	2	
Love Child II.. .	H.	147	3	5	
Loved One (By) .	G.	170	8	A.	
——— . .	C.	170	8	2	
——— . .	F.	170	8	2	
Love Drift . .	M.	794	2	5	
Loveite . . .	F.	526	4	3	
Love Jack . .	G.	134	8	4	Jack, 24=6.
Love Leaf . .	M.	230	5	5	Leaf, 120=3.
Love Lost . .	M.	602	8	6	
Love Philtre . .	F.	820	1	2	
Love Quest . .	F.	600	6	3	
Love Wisely (By) .	F.	163	1	2	
Lowland Aggie .	F.	162	9	2	Aggie, 41=5.
Lucerne . . .	G.	340	7	6	
Lucinda . . .	F.	145	1	4	
Lucknow . . .	H.	111	3	A.	
Lucky George .	G.	268	7	A.	George, 208=1.
Lucky Jim . .	G.	103	4	5	Jim, 43=7.
Lucky John . .	G.	115	7	5	John, 55=1.
Luculia . . .	F.	91	1	2	
Lucy Bertram .	F.	948	3	2	Lucy, 106=7.
Luerana . . .	F.	292	4	4	
Luff-a-Lee . .	F.	151	7	3	
Luisant . . .	C.	141	6	3	
Luke Ward . .	C.	267	6	4	
Lurgan . . .	G.	300	3	6	
Lusignan . . .		210	3	3	
Lustroso . . .	G.	762	6	6	
Lutin . . .	G.	480	3	3	
Lychnobite . .	G.	491	5	3	
Lyddite . . .	G.	434	2	5	
Lye Less . . .	F.	140	5	4	
Lyndon Green .	G.	414	9	4	Green, 280=1.
Lynton . . .	C.	530	8	2	
Lyonstown . .	M.	606	3	A.	

Name.	Sex.	Value.	Remainder.	Age (figures indicate years and " A " aged).	Remarks.
Macedoine . .	F.	171	9	2	
M'Callum Mare .	H.	376	7	5	M'Callum, 130 = 4.
Maccoon . . .	G.	116	8	3	
Macheath (By) .	F.	481	4	2	
M'Mahon II. . .	G.	151	7	A.	
M'Mayne . . .	C.	160	7	4	
M'Yardley . .	C.	315	9	4	
Machine . . .	M.	400	4	A.	
Maciso . . .	C.	167	5	3	
Macready's Hope .	C.	383	5	4	Hope, 91 = 1.
Madam II. . .	M.	85	4	A.	
Madame Danglass .	M.	450	9	6	Danglass, 365 = 5.
Madame de Montespan . . .	F.	792	9	2	Montespan, 693 = 9.
Maddington . .	C.	565	7	3	
Mademoiselle de Arcizac . . .	M.	431	8	5	Arcizac, 289 = 1.
Mademoiselle de Longchamps .	M.	438	6	5	Longchamps, 286 = 7.
Madge Ford .	F.	334	1	3	Madge, 44 = 8.
Mad Lorrimer . .	G.	517	4	A.	Lorrimer, 472 = 4.
Madrid . . .	F.	259	7	2	
Mafra . . .	F.	322	7	2	
Magic Box . .	G.	148	4	6	Box, 84 = 3.
Magic Lantern .	G.	795	3	A.	Lantern, 731 = 2.
Magic Mirror . .	G.	504	9	4	Mirror, 440 = 8.
Magnate . . .	G.	521	8	A.	
Magnesia . . .	F.	192	3	2	
Magnus . . .	C.	171	9	3	
Mahalie . . .	G.	86	5	4	
Mahratta . . .	C.	646	7	4	
Maidenhair . .	M.	319	4	6	
Maid of Clwyd .	F.	195	6	2	Maid, 54 = 9.
Maid of Oak .	F.	162	9	4	Maid, 54 = 9.
Maid of the Forest .	F.	887	5	4	Maid, 54 = 9.
Maid of Valetta .	M.	265	4	A.	Maid, 54 = 9.
Mail	G.	80	8	4	
Mailed Fist . .	C.	624	3	2	Fist, 540 = 9.
Maine Mor . .	H.	342	9	5	Maine, 100 = 1.
Main Point . .	G.	632	2	A.	Point, 532 = 1.
Mainspring . .	G.	510	6	A.	
Main Top . .	C.	582	6	2	Top, 482 = 5.
Main Yard . .	G.	315	9	6	Yard, 215 = 8.

Name.	Sex.	Value.	Remainder.	Age (figures indicate years and "A" aged).	Remarks.
Maisee II. . .	F.	120	3	4	
Maitland . . .	G.	535	4	4	
Majestic II. . .	G.	534	3	A.	
Major . . .	G.	244	1	5	
Major II. . . .	G.	244	1	4	
Malachite . .	G.	502	7	A.	
Malatesta . .	F.	943	7	4	
Malcolm Orme .	C.	373	4	2	Malcolm, 131 = 5.
Malde Mer . .	M.	325	1	5	Mal, 71 = 8.
Mallard . . .	G.	275	5	6	
Mallerstrang .	C.	1002	3	3	
Maltese Cross .	G.	770	5	A.	Cross, 282 = 3.
Malva . . .	F.	152	8	2	
Malvern Hill .	G.	436	4	6	Hill, 35 = 8.
Mambrino . .	H.	339	6	5	
Manager . .	G.	295	7	A.	
Manatee . .	G.	502	7	6	
Mandelay . .	C.	145	1	4	
Mango Relish .	F.	657	9	4	Relish, 540 = 9.
Manhattan Boy .	G.	559	1	4	Boy, 14 = 5.
Manifesto . .	G.	256	4	A.	
Manna . . .	G.	297	9	4	
Mannikin . .	G.	161	8	5	
Mannlicher . .	C.	341	8	4	
Manston . .	H.	601	7	6	
Mantilla II. . .	M.	522	9	5	
Manulla . .	G.	122	5	5	
Manx Lad . .	G.	206	8	3	Lad, 35 = 8.
Manx Penny .	G.	321	6	4	Penny, 150 = 6.
Maori Chieftain .	C.	800	8	3	Chieftain, 543 = 3.
Maori Queen II. .	M.	337	4	A.	Queen, 80 = 8.
Marauder . .	G.	450	9	6	
Marceline . .	M.	391	4	6	
Marcellus . .	G.	401	5	A.	
Marcha Real .	M.	485	8	6	March, 245 = 2 ; Real, 240 = 6.
Marchmont . .	C.	336	3	2	
Marcite . . .	F.	701	8	2	
Marco (By) . .	F.	267	6	2	
——— . . .	F.	267	6	2	
——— . . .	C.	267	6	2	
Marconi . . .	H.	327	3	5	
Mardi Gras . .	C.	476	8	4	Mardi, 255 = 3.
Mardonius . .	C.	371	2	4	

Name.	Sex.	Value.	Remainder.	Age (figures indicate years and "A" aged).	Remarks.
Marechal Niel . . .	C.	391	4	3	Marechal, 301 = 4.
Mareen . . .	M.	301	4	6	
Marengo . . .	C.	327	3	3	
Maresco . . .	G.	337	4	5	
Margaret . . .	F.	871	7	2	
Margarita M. . .	M.	923	5	6	Margarita, 872 = 8.
Margo . . .	F.	267	6	3	
Marialva . . .	M.	353	2	5	
Marie S. . .	F.	322	7	4	Marie, 251 = 8.
Marievale . . .	F.	371	2	3	
Marigold II. . .	M.	301	4	A.	
Mark Tapley . .	C.	782	8	3	Mark, 261 = 9.
Marmelo . . .	C.	327	3	3	
Marmion . . .	C.	341	8	2	
Marmiton (By) .	F.	731	2	2	
——— . .	C.	731	2	2	
Marmont . . .	G.	333	9	A.	
Marmor . . .	C.	481	4	3	
Marmot . . .	C.	287	8	3	
Marmouts . . .	C.	747	9	4	
Marouette . .	F.	657	9	3	
Marpessa . .	G.	392	5	5	
Marriage Lines .	F.	384	6	4	
Marsaba . . .	H.	314	8	5	
Marsala II. .	M.	333	9	5	
Marsden Rock .	G.	577	1	A.	Rock, 222 = 6.
Marshal . . .	G.	571	4	A.	
Marchcress . .	F.	831	3	4	
Martagon (By) .	F.	712	1	2	
——— . .	G.	712	1	2	
Marteau . . .	C.	647	8	4	
Marten . . .	G.	691	7	5	
Marthus . . .	G.	706	4	A.	
Martinez del Rio .	C.	961	7	3	
Martinmas . .	G.	792	9	2	
Marty . . .	G.	651	3	A.	
Martyrdom . .	G.	885	3	4	
Marvel II. . .	G.	361	1	A.	
Mary Anderson .	M.	626	5	5	Mary, 251 = 8.
Marybridge . .	F.	456	6	2	Mary, 251 = 8.
Mary Elizabeth .	M.	707	5	A.	Mary, 251 = 8.
Mary Josephine .	F.	407	2	4	Mary, 251 = 8.
Mascagni . . .	H.	201	3	5	
Masher . . .	C.	541	3	5	

Name.	Sex.	Value.	Remainder.	Age (figures indicate years and "A" aged).	Remarks.
Masonic Jewel . .	M.	221	5	5	Jewel, 49＝4.
Masquerade . .	F.	335	2	4	
Massachusetts . .	C.	641	2	3	
Masserene . .	H.	361	1	A.	
Master II. . .	G.	701	8	5	
Master Charley .	C.	945	9	2	Charlie, 244＝1.
Master Gamp . .	G.	842	5	5	Gamp, 141＝6.
Master Harry . .	G.	917	8	6	Harry, 216＝9.
Master Herbert .	G.	1508	5	5	Herbert, 807＝6.
Master Hugh . .	G.	722	2	6	Hugh, 21＝3.
Master Johnny . .	G.	766	1	5	Johnny, 65＝2.
Master Lovat . .	C.	1218	3	4	Lovat, 517＝4.
Master Ludwig .	C.	761	5	3	Ludwig, 60＝6.
Master Morglay .	G.	1007	8	5	Morglay, 306＝9.
Master Newby . .	C.	779	5	2	Newby, 78＝6.
Master Orme . .	G.	943	7	5	Orme, 242＝8.
Master Osmunda .	C.	858	3	2	Osmunda, 157＝4.
Master Tom . .	G.	1143	9	4	Tom, 442＝1.
Master Trillion .	C.	1391	5	2	Trillion, 690＝6.
Master Victor . .	C.	1401	6	2	Victor, 700＝7.
Master Willie . .	H.	747	9	6	Willie, 46＝1.
Matagon . . .	F.	512	8	4	
Match . . .	F.	44	8	4	
Matchbelle . .	F.	86	5	2	
Match Boy . .	C.	58	4	3	Boy, 14＝5.
Matchmaker (By) .	C.	305	8	3	
——— . . .	F.	305	8	2	
——— . . .	F.	305	8	2	
Matchman . .	G.	135	9	4	
Matchworks . .	C.	330	6	4	
Matfen . . .	G.	582	6	A.	
Mathew . . .	G.	462	3	6	
Mathioli . . .	G.	502	7	A.	
Mat Sallep . .	C.	570	3	4	Salleh, 129＝3.
Maudlin . . .	M.	126	9	5	
Maude's Pride . .	M.	337	4	6	
Maund . . .	F.	96	6	4	
Mauvezin . . .	H.	194	5	6	
Mavronero . .	C.	593	8	3	
Maximum . .	C.	201	3	3	
May Boy II. . .	G.	64	1	A.	Boy, 14＝5.
May Day . . .	G.	64	1	5	Boy, 14＝5 ; May, 50＝5.
May Fitz . . .	M.	537	6	5	May, 50＝5.

Name.	Sex.	Value.	Remainder.	Age (figures indicate years and "A" aged).	Remarks.
Mayflower	F.	368	8	2	
May Flower II.	M.	368	8	A.	
Mayfly	G.	170	8	4	
Mayfly	G.	170	8	6	
May Hastings	M.	593	8	6	May, 50 = 5.
May I	M.	61	7	5	
May King	G.	140	5	6	May, 50 = 5.
May Lassie	M.	151	7	A.	May, 50 = 5.
May Morning	F.	412	7	4	May, 50 = 5.
Mayo's Pride	G.	347	5	A.	Pride, 284 = 5.
May's Pride	G.	341	8	4	Pride, 284 = 5.
Maythorne II.	G.	707	5	A.	
Mazawattee	G.	74	2	6	
M. C.	G.	121	4	4	
M. D.	G.	65	2	A.	
Mead	C.	54	9	2	
Meadow	F.	60	6	4	
Meadow Daisy	M.	91	1	5	Daisy, 31 = 4.
Meadow Dancer	G.	375	6	A.	Dancer, 315 = 9.
Meadow Fescue	M.	246	3	A.	Fescue, 186 = 6.
Meadow Gold	G.	120	3	6	Gold, 60 = 6.
Meadow Stream	M.	770	5	A.	Stream, 710 = 8.
Mealy	G.	90	9	5	
Measles		200	2	6	
Medal	F.	84	3	4	
Medallion	F.	145	1	4	
Meditation	F.	805	4	2	
Medmenham (By)	F.	199	1	3	
Medor	C.	256	4	2	
Meenan	G.	150	6	A.	
Melba	M.	83	2	A.	
Meldhre	F.	299	2	3	
Melete	M.	490	4	5	
Meliboea	F.	93	3	4	
Melito	G.	95	5	6	
Mellin	G.	130	4	6	
Melsary	C.	350	8	2	
Melton (By)	C.	530	8	2	
——	C.	530	8	2	
——	C.	530	8	2	
——	C.	530	8	2	
——	F.	530	8	2	
——	C.	530	8	2	
Memoria	M.	307	1	6	

Name.	Sex.	Value.	Remainder.	Age (figures indicate years and "A" aged).	Remarks.
Menander . .	C.	355	4	4	
Mender . .	G.	304	3	5	
Menelik . .	G.	160	7	6	
Menteith . .	H.	915	6	6	
Merchant Prince .	G.	1083	3	6	Prince, 390=3.
Mercury II. .	H.	470	2	A.	
Mereclough .	M.	306	9	5	
Mere Skipper .	C.	532	1	3	
Merit . .	G.	650	2	6	
Mermaid . .	F.	294	6	4	
Merriment .	F.	750	3	3	
Merrion . .	C.	310	4	4	
Merry Agnes .	F.	401	5	2	Agnes, 141=6.
Merry Andrew .	G.	531	9	3	Andrew, 271=1.
Merry Bout .	F.	668	2	4	Bout, 408=3.
Merry Go Round .	C.	546	6	4	Go Round, 286=7.
Merry Go Round (By)	H.	546	6	5	Go Round, 286=7.
Merry Knight .	G.	710	8	5	Knight, 450=9.
Merry Love .	M.	370	1	A.	Love, 110=2.
Merry Maiden .	M.	364	4	5	Maiden, 104=5.
Merry Marie .	M.	511	7	9	Marie, 251=8.
Merry Methodist	H.	1179	9	6	Methodist, 919=1.
Merry Monk II. .	G.	372	3	A.	Monk, 112=4.
Merry Mood .	G.	310	4	A.	Mood, 50=5.
Merry Prince .	G.	650	2	4	Prince, 390=3.
Merry Shields .	M.	664	7	6	Shields, 404=8.
Merry Tar .	C.	861	6	3	Tar, 601=7.
Merry Wink .	C.	336	3	4	Wink, 76=4.
Merveilleuse .	M.	383	5	5	
Merville . .	M.	360	9	6	
Mesalliance .	F.	261	9	2	
Messene . .	H.	170	8	6	
Metal Man .	G.	571	4	A.	Man, 91=1.
Methelios .	G.	567	9	5	
Method . .	G.	459	9	6	
Michael . .	G.	101	2	6	
Microbe . .	C.	268	7	3	
Microbe . .	C.	268	7	4	
Middleham (By) .	G.	119	2	3	
Midnight . .	M.	494	8	A.	
Midnight II. .	G.	494	8	A.	
Midnight Mass .	F.	595	1	4	Mass, 101=2.
Miff . .	F.	120	3	2	
Mighty King .	H.	540	9	A.	King, 90=9.

Name.	Sex.	Value.	Remainder.	Age (figures indicate years and "A." aged).	Remarks.
Milady . . .	F.	85	4	4	
Milano . . .	G.	127	1	4	
Milesman . .	G.	220	4	5	
Miliolite . .	M.	516	3	5	
Milk Boy . .	G.	104	5	A.	Boy, 14=5.
Millbury . .	M.	282	3	5	
Millepede . .	G.	174	3	A.	
Millpark . .	G.	371	2	A.	
Milner (By) .	C.	320	5	2	
—— . .	C.	320	5	3	
Miltiades . .	H.	502	7	A.	
Milton Chapel II.	C.	634	4	4	Chapel, 114=6.
Mimosa San .	M.	258	6	5	
Min . . .	M.	90	9	6	
Mindoro . .	C.	306	9	3	
Miner . .	G.	290	2	A.	
Mingle . .	G.	140	5	5	
Minicoy . .	G.	122	5	5	
Miniford . .	C.	374	5	3	
Minima . .	F.	131	5	3	
Ministre . .	H.	750	3	5	
Minor Daly .	C.	335	2	4	Daly, 45=9.
Mi Novia . .	M.	197	8	6	Novia, 147=3.
Minskip . .	H.	250	7	5	
Minstead . .	C.	564	6	3	
Minstrel Girl II.	M.	1030	4	6	Girl, 250=7.
Mintberry .	F.	712	1	2	
Mintseed . .	C.	564	6	3	
Mintstalk . .	G.	972	9	6	
Minx . .	F.	170	8	2	
Mirobolant .	C.	335	2	4	
Mirthful (By) .	C.	755	8	2	
Mirus . .	C.	300	3	4	
Mischief . .	F.	183	3	4	
Misdeal . .	G.	144	9	A.	
Miss Avis .	M.	241	7	5	
Miss Bee . .	F.	112	4	3	Bee, 12=3.
Miss Bentley .	M.	602	8	A.	Bentley, 502=7.
Miss Bertha .		708	6	A.	Bertha, 608=5.
Miss Blossom .	F.	234	9	3	Blossom, 134=8.
Miss Bobs .	F.	166	4	3	Bobs, 156=3.
Miss Brice .	F.	362	2	2	Brice, 262=1.
Miss Bryant .	F.	752	5	3	Bryant, 652=4.
Miss Cheriton .	F.	763	7	2	Cheriton, 663=6.

Name.	Sex.	Value.	Remainder.	Age (figures indicate years and "A" aged).	Remarks.
Miss Clifden II. .	M.	284	5	6	Clifden, 184=4.
Miss Cronkhill. .	F.	427	4	4	Cronkhill, 327=3.
Miss Darlow . .	F.	341	8	3	Darlow, 241=7.
Miss De Wet . .	M.	530	8	5	Wet, 416=2.
Miss Diana . .	M.	166	4	5	Diana, 156=3.
Miss Doods . .	F.	174	3	4	Doods, 74=2.
Miss Doris . .	F.	366	6	2	Doris, 266=5.
Miss Drake . .	F.	334	1	2	Drake, 234=9.
Missey's Pet . .	F.	607	4	4	Pet, 490=4.
Miss Garnet . .	F.	781	7	2	Garnet, 681=6.
Miss Gladys . .	F.	215	8	3	Gladys, 115=7.
Miss Grab . .	M.	323	8	5	Grab, 223=7.
Miss Heather . .	M.	319	4	6	Heather, 219=3.
Miss Hobhouse .	F.	180	9	4	Hobhouse, 80=8.
Miss Hugo . .	F.	137	2	4	Hugo, 37=1.
Miss Hugoin . .	F.	177	6	2	Hugoin, 77=5.
Missionary . .	G.	600	6	A.	
Miss Jason . .	F.	214	7	4	Jason, 114=6.
Miss Lettice . .	F.	209	2	4	Lettice, 109=1.
Miss Mermaid. .	M.	394	7	6	Mermaid, 294=6.
Miss Morgan II. .	M.	412	7	6	Morgan, 312=6.
Miss Morrison. .	M.	452	2	A.	Morrison, 352=1.
Miss Muriel II. .	M.	396	9	6	Muriel, 296=8.
Miss Mustard . .	F.	804	3	2	Mustard, 704=2.
Miss Pac . . .	F.	201	3	4	Pac, 101=2.
Miss Puff . . .	M.	260	8	A.	Puff, 160=7.
Miss Puss . . .	F.	240	6	4	Puss, 140=5.
Miss Riley . .	F.	340	7	3	Riley, 140=5.
Miss Rosamond .	M.	408	3	A.	Rosamond, 308=2.
Miss Royston . .	M.	822	3	A.	Royston, 722=2.
Miss Tat . . .	M.	901	1	A.	Tat, 801=9.
Miss Tessie . .	F.	580	4	2	Tessie, 480=3.
Miss Unicorn . .	M.	448	7	6	Unicorn, 348=6.
Miss Villiers . .	F.	480	3	2	Villiers, 380=2.
Miss Walkaway .	M.	144	9	6	Walkaway, 44=8.
Mr. Brown . .	H.	358	7	6	Brown, 258=6.
Mr. Dunlop . .	G.	266	5	A.	Dunlop, 166=4.
Mr. Morley (By) .	M.	986	5	5	Morley, 286=7.
Mr. Quilp . .	G.	830	2	5	Quilp, 130=4.
Mr. Schomberg .	H.	1268	8	5	Schomberg, 568=1.
Mr. Tullock . .	G.	1150	7	3	Tullock, 450=9.
Mrs. Gamp . .	C.	248	5	2	Gamp, 141=6.
Mrs. Har . . .	F.	313	7	3	Har, 206=8.
Mrs. Leo . . .	M.	153	9	6	Leo, 46=1.

Name.	Sex.	Value.	Remainder.	Age (figures indicate years and "A" aged).	Remarks.
Mister . . .	C.	501	6	2	
Misterman . .	G.	790	7	6	
Mistress . . .	M.	770	5	6	
Misty Love . .	C.	620	8	2	Love, 110=2.
Misunderstood. .	M.	819	9	6	
Mittimus . . .	M.	149	5	6	
Mixed Powder .	C.	442	1	2	Powder, 290=2.
Mobility . . .	F.	484	7	3	
Mocanna (By) .	F.	118	1	2	
Model . . .	G.	86	5	A.	
Moderate. . .	C.	656	8	4	
Modest Moll .	M.	588	3	6	Moll, 72=9.
Molester . .	C.	746	8	4	
Molly . . .	F.	82	1	4	
Molly Mawk .	F.	144	9	3	Molly, 82=1.
Monachus . .	G.	177	6	A.	
Monaghan . .	G.	167	5	A.	
Monarch . .	H.	312	7	6	
Mona's Queen .	F.	184	4	4	Queen, 80=8.
Mondaine . .	M.	156	3	A.	
Mondella . .	F.	137	2	6	
Monedula. . .	M.	141	6	5	
Moneyspinner .	C.	490	4	4	
Monitress . .	F.	762	6	3	
Monk II. . . .	G.	112	4	A.	
Monkland . .	H.	197	8	6	
Monksilver . .	G.	482	5	A.	
Monkspath . .	G.	658	1	6	
Monkwood . .	G.	122	5	6	
Monmouth . .	H.	537	6	6	
Monoeci Arx .	C.	457	7	4	
Monotype . .	G.	578	2	6	
Monster . . .	G.	752	5	6	
Mont de Piete .	H.	606	3	A.	Mont, 92=2.
Montella . . .	F.	533	2	3	
Montgaillard II. .	C.	757	1	3	
Monxton . . .	C.	622	1	2	
Moondyne II. . .	G.	160	7	A.	
Moonlit . . .	G.	526	4	6	
Moon Ray . .	F.	306	9	4	
Moonstone II.. .	F.	612	9	4	
Moonsprite . .	F.	986	5	4	
Mopish . . .	M.	422	8	5	
Moraine . . .	M.	306	9	6	

Name.	Sex.	Value.	Remainder.	Age (figures indicate years and "A" aged).	Remarks.
Moral Mary	M.	523	1	A.	
Morass	F.	307	1	2	
Mordant	G.	697	4	6	
Morecambe Bay	C.	318	3	3	Bay, 12 = 3.
Morelos	F.	348	6	3	
More Power	G.	534	3	A.	Power, 288 = 9.
Morganatic	C.	734	5	3	
Morgante	G.	713	2	6	
Morglette	M.	702	9	A.	
Morion (By)	F.	306	9	2	
Morningdew	G.	382	4	A.	
Morning Glass	C.	473	5	3	Glass, 111 = 3.
Morning Gun	H.	432	9	6	Gun, 70 = 7.
Mornock	G.	314	8	6	
Morocco Bound	C.	330	6	3	
Morpheus	C.	392	5	2	
Mort Eau	F.	649	1	3	
Morville	C.	362	2	4	
Moss Agate	F.	533	2	2	
Most Excellent	H.	1087	7	5	Excellent, 581 = 5.
Mount	G.	496	1	A.	
Mountain Air	G.	757	1	A.	Air, 211 = 4.
Mountain Buck	H.	568	1	5	Buck, 22 = 4.
Mountain Daisy	F.	577	1	3	Daisy, 31 = 4.
Mountain Maid	M.	600	6	6	Maid, 54 = 9.
Mountain Mist	C.	1046	2	3	Mist, 500 = 5.
Mountain Rose	C.	759	3	2	Rose, 213 = 6.
Mountains High	F.	561	3	2	High, 15 = 6.
Mount Cashel	G.	847	1	A.	Cashel, 351 = 9.
Mount Dalton	G.	981	9	A.	Dalton, 485 = 8.
Mount Hawk	G.	523	1	A.	Hawk, 27 = 9.
Mount Judkin	C.	573	6	3	Judkin, 77 = 5.
Mount Keen	C.	576	9	4	Keen, 80 = 8.
Mount Lyell	G.	567	9	4	Lyell, 71 = 8.
Mount Prospect	H.	1348	7	A.	Prospect, 852 = 6.
Mount Thomas	C.	998	8	3	Thomas, 502 = 7.
Mousquetaire (By)	C.	746	8	2	
Mousse	F.	116	8	2	
Moyfemath	M.	798	6	A.	
Moyola	F.	83	2	4	
Much Too Early	C.	690	6	4	Early, 241 = 7.
Muckross	G.	322	7	4	
Muff	F.	120	3	2	
Mug	G.	60	6	2	

Name.	Sex.	Value.	Remainder.	Age (figures indicate years and "A" aged).	Remarks.
Muggins II.	G.	170	8	6	
Mulatto	G.	86	5	A.	
Mulhuddert		679	4	6	
Mummy	M.	90	9	5	
Mundon	F.	144	9	3	
Munn	G.	90	9	5	
Muirland Witch	M.	350	8	6	Witch, 9=9.
Murad	G.	245	2	5	
Murillo	G.	286	7	A	
Murlingden	F.	394	7	3	
Mummur	F.	480	3	2	
Musette	F.	457	7	2	
Musicwood	M.	93	3	5	
Musket Stock	H.	1012	4	5	Stock, 482=5.
Muskham	H.	165	3	A.	
Mustapha	G.	1231	7	2	Mustapha, 229=4.
Mustard II.	G.	704	2	A.	
My Lollypop	F.	284	5	3	Lollypop, 234=9.
My Mascotte	F.	573	6	2	Mascotte, 523=1.
Mynetian	C.	560	2	2	
Myra Hamilton	M.	767	2	5	Myra, 241=7.
Myrcia	F.	311	5	4	
My Rose	M.	263	2	5	
Myrtleberry	H.	882	9	6	
Mysterious Lady	M.	825	6	6	Lady, 45=9.
Mystic Maid	F.	574	7	2	Maid, 54=9.
Mystic Moon	C.	616	4	4	Moon, 96=6.
Naas	G.	112	4	3	
Nahlband	C.	206	8	4	
Nah'ma'wush	H.	178	7	A.	
Naivete	C.	550	1	3	
Nakhillak	H.	691	7	6	
Nameless Girl	F.	450	9	3	Girl, 250=7.
Nancy Dawson	M.	287	8	A.	Nancy, 171=9.
Nancy Lee II.	M.	211	4	3	Nancy, 171=9.
Nansen	H.	211	4	A.	
Naomi	M.	107	8	6	
Nappaby	G.	66	3	6	
Napper Tandy	G.	718	7	5	
Narcissus	C.	491	5	3	
Narcissus II.	G.	491	5	A.	
Naseby	G.	79	7	A.	
Nata	M.	452	2	A.	

Name.	Sex.	Value.	Remainder.	Age (figures indicate years and "A" aged).	Remarks.
Natalina . . .	F.	533	2	3	
Nat Gould . .	G.	511	7	A.	Nat, 451 = 1.
Naughty Chat . .	F.	866	2	3	Chat, 404 = 8.
Nebelig . . .	C.	122	5	2	
Necromancer (By) .	F.	637	7	2	
Nectarine . .	M.	740	2	5	
Ned Lyons . .	C.	214	7	3	Ned, 64 = 1.
Ned o' the Hills .	G.	180	9	A.	Ned, 64 = 1.
Neish . . .	H.	360	9	A.	
Nell	M.	90	9	6	
Nelly Clare . .	F.	360	9	4	Nelly, 100 = 1.
Nelson III. . .	G.	200	2	A.	
Nenemoo'sha .	M.	467	8	6	
Neptuna . . .	F.	200	2	2	
Nereus . . .	G.	330	6	4	
Nervous . . .	M.	390	3	5	
Nestor . . .	G.	720	9	4	
Netherland . .	C.	349	7	4	
Nethersted . .	H.	738	9	5	
Nether Wallop .	G.	381	3	5	
Nettie . . .	F.	470	2	3	
Nettlecreeper .	M.	1000	1	5	
Never Say Never .	C.	750	3	3	
New Antigone .	F.	599	5	2	Antigone, 533 = 2.
New Barns (By) .	G.	379	1	3	Barns, 313 = 7.
Newbridge . .	F.	271	1	4	Bridge, 205 = 7.
New Broom . .	M.	314	8	5	Broom. 248 = 5.
Newbury . . .	G.	278	8	A.	
New Century . .	C.	399	3	2	Century, 333 = 9.
New Century Rose .	F.	612	9	3	Rose, 213 = 6.
New Fashion . .	F.	497	2	4	Fashion, 431 = 8.
New Fashion . .		497	2	A.	Fashion, 431 = 8.
New Jersey . .	G.	339	6	A.	Jersey, 273 = 3.
New Jersey II. .	C.	339	6	3	Jersey, 273 = 3.
New Norfolk . .	C.	418	4	4	Norfolk, 352 = 1.
New Palace . .	C.	237	3	2	Palace, 171 = 9.
Newton Lad . .	G.	551	2	6	Lad, 35 = 8.
Newtown . . .	C.	522	9	4	
New York II. . .	C.	298	1	3	York, 232 = 7.
Niam . . .	M.	101	2	A.	
Nigger III. . .	G.	270	9	6	
Nightjar . . .	G.	654	6	A.	
Nightjar II. . .	G.	654	6	A.	
Nightshade . .	G.	764	8	5	

Name.	Sex.	Value.	Remainder.	Age (figures indicate years and "A" aged).	Remarks.
Night-time	M.	890	8	5	
Night Wanderer	G.	911	2	5	Wanderer, 461 = 2.
Nil Desperandam	H.	549	9	A.	
Nimble Girl	M.	372	3	6	Girl, 250 = 7.
Nimrod II.	G.	300	3	A.	
Ninon	M.	150	6	5	
Nipperkin	C.	322	7	4	
Nippon	H.	102	3	A.	
Nirvana	H.	382	4	5	
Nitrate Maid	F.	1114	7	4	Maid, 54 = 9.
Nivernais	C.	390	3	3	
No	H.	56	2	A.	
Noble Heart	H.	694	1	6	Heart, 606 = 3.
Nobleman II.	G.	178	7	A.	
Noblethorpe	C.	775	1	3	
Nobold	H	98	8	A.	
Nochty	C.	482	5	3	
No Class	F.	167	5	4	
No Credit	F.	690	6	4	
No Denial	F.	150	6	2	
No Fear	F.	346	4	4	
No Fool	G.	172	1	A.	
No Lady	F.	101	2	3	
Nolo	G.	92	2	5	
Nomad	G.	101	2	5	
None so Pretty	M.	475	7	5	
Nonna	F.	107	8	4	
Nono	F.	112	4	4	
Noonday II.	F.	120	3	4	
Noorong	G.	328	4	4	
Nora III.	M.	257	5	A.	
Nora IV.	M.	257	5	A.	
Nora Mea	M.	308	2	6	
Nordrach	C.	277	7	4	
Nor' Easter	G.	923	5	6	Easter, 671 = 5.
Norham	G.	297	9	2	
Norman II.	G.	342	9	A.	
Normanby	C.	354	3	4	
Norse Girl	F.	562	4	2	Girl, 250 = 7.
Norseman II.	G.	402	6	4	
North Crawley	G.	919	1	5	Crawley, 262 = 1.
Northern Farmer	H.	1027	1	A.	Farmer, 521 = 8.
Northern Light II.	G.	936	9	A.	Light, 430 = 7.
North Pole	G.	773	8	3	Pole, 116 = 8.

Name.	Sex.	Value.	Remainder.	Age (figures indicate years and "A" aged).	Remarks.
North Sea . .	H.	727	7	A.	Sea, 70=7.
North Tyne . .	H.	1107	9	5	Tyne, 450=9.
Nosbine . . .	C.	174	3	4	
Nota Bene . .	M.	529	7	A.	
Not Guilty . .	G.	912	3	3	
Nothing II. . .	G.	525	3	A.	
Not Out . . .	C.	859	4	4	
Notre Mere . .	F.	906	6	4	
Nougat . . .	M.	477	9	5	
Novelty . . .	G.	572	5	6	
Novice II. . .	G.	192	3	A.	
Novice III. . .	M.	192	3	5	
Now. . . .	G.	56	2	6	
Nuala . . .	M.	88	7	6	
Nunnykirk . .	G.	350	8	A.	
Oasis . . .	G.	127	1	5	
Oban . . .	G.	60	6	A.	
Obelisk II. . .	M.	129	3	5	
Ocean Blue . .	G.	405	9	5	Blue, 38=2.
Ocean Nymph .	F.	537	6	3	Nymph, 170=8.
Oceano . . .	M.	374	5	A.	
Ocean Rover .	H.	853	7	A.	Rover, 486=9.
O'Connell II. .	G.	119	2	A.	
Octoroon Girl .	F.	928	1	4	Girl, 250=7.
Odawara . .	F.	220	4	3	
Oddfish . .	G.	386	8	4	
O'Donovan Rossa .	G.	456	6	5	O'Donovan, 193=4.
Odoratus . .	G.	678	3	3	
Odran . . .	G.	262	1	5	
Oedipus . .	G.	161	8	A.	
Oenopion . .	H.	213	6	6	
Offertory . .	G.	892	1	6	
Oh Bang . .	G.	80	8	2	
Ohio . . .	H.	28	1	5	
Oh My . .	G.	57	3	A.	
Okapi . . .	C.	118	1	2	
Old Ale . .	G.	82	1	A.	Ale, 41=5.
Old Bess . .	M.	113	5	5	Bess, 72=9.
Old Fashioned .	F.	476	8	4	Fashioned, 435=3.
Oldham . .	G.	86	5	4	
Old Head . .	C.	60	6	4	Head, 19=1.
Old John . .	G.	96	6	5	John, 55=1.
Old Pal . .	G.	152	8	6	Pal, 111=3.

Name.	Sex.	Value.	Remainder.	Age (figures indicate years and "A" aged).	Remarks.
Old Patch . .	C.	125	8	4	Patch, 84=3.
Old Sol . . .	C.	133	7	3	Sol, 92=2.
Old Taylor . .	C.	681	6	4	Taylor, 640=1.
Old Tim . . .	G.	481	4	5	Tim, 440=8.
Old Times . .	M.	488	2	A.	Times, 447=6.
Old Times . .	C.	488	2	2	Times, 447=6.
Oldtown . . .	H.	497	2	A.	
Old Windsor . .	H.	308	2	6	Windsor, 267=6.
Olea	F.	48	3	4	
Oleaster . . .	F.	708	6	4	
Olibanum . .	G.	126	9	A.	
Olivarez . . .	C.	330	6	3	
Olive II. . .	M.	112	4	A.	
Olive Branch . .	G.	368	8	A.	Branch, 256=4.
Olive Branch II. .	H.	368	8	5	Branch, 256=4.
Olivette . . .	M.	522	9	A.	
Olivia . . .	F.	123	6	2	
Ollav . . .	G.	118	1	5	
Olympian . .	C.	217	1	·4	
Omaha II. . .	C.	54	9	3	
Omar Khayyam .	C.	961	7	2	Omar, 310=4.
Omelet . . .	C.	482	5	4	
Omelettina . .	F.	543	3	2	
Ondulee . . .	F.	96	6	4	
One	G.	56	2	4	
One and All . .	F.	143	8	3	One, 56=2.
One D . . .	F.	70	7	2	D, 14=5.
Only Child . .	F.	134	8	2	Child, 37=1.
Only Your Uncle .	H.	414	9	5	Uncle, 101=2.
Oof Bird . . .	F.	293	5	3	Bird, 206=8.
Oporto . . .	M.	694	1	A.	
Oppressor . .	H.	474	6	6	
Opsal . . .	G.	173	2	6	
Opulence . . .	F.	239	5	3	
Orangeman . .	G.	345	3	3	
Orange Pat . .	G.	736	7	6	Pat, 481=4.
Orange Pip . .	G.	415	1	A.	Pip, 160=7.
Orbay . . .	C.	214	7	2	
Orbel . . .	C.	234	9	4	
Orchid . . .	C.	209	2	4	
Orestes . . .	H.	689	5	A.	
Or Ever . . .	H.	493	7	6	Ever, 291=3.
Orfio . . .	G.	298	1	6	
Organ Grinder .	G.	746	8	A.	Grinder, 474=6.

Name.	Sex.	Value.	Remainder.	Age (figures indicate years and "A" aged).	Remarks.
Orillon . . .	C.	282	3	2	
Orion (By) . .	C.	267	6	2	
—— . . .	F.	267	6	3	
—— . . .	F.	267	6	2	
—— . . .	G.	267	6	3	
—— . . .	F.	267	6	2	
—— . . .	C.	267	6	2	
Oriole . . .	G.	253	1	A.	
Ormac (By) . .	C.	262	1	2	
Orme (By) . .	F.	242	8	3	
—— . . .	F.	242	8	2	
Ormeau . . .	H.	248	5	6	
Ormelie . . .	C.	292	4	3	
Ormenus . . .	C.	362	2	4	
Ormes Bay . .	F.	314	8	3	
Orontes (By) . .	C.	676	1	3	
Oroide . . .	M.	227	2	6	
Oroya . . .	C.	224	8	2	
Orphrey . . .	F.	492	6	3	
Orpington . .	G.	802	1	6	
Orrag . . .	C.	228	3	2	
Orris Root . .	H.	868	4	A.	Root, 606 = 3.
Orsay . . .	G.	272	2	6	
Ortelle . . .	F.	642	3	3	
Ortygian . . .	G.	665	8	5	
Orvietan . . .	G.	742	4	3	
Oryx . . .	M.	282	3	6	
Osbech . . .	G.	94	4	A.	
Osboch . . .	C.	86	5	4	
Osborne . . .	G.	314	8	5	
Osy . . .	F.	77	5	3	
Osyth . . .	F.	472	4	2	
Othello II. . .		458	8	A.	
Othery . . .	M.	617	5	A.	
Ottoman II. . .	G.	661	4	6	
Oughterard . .	G.	817	7	A.	
Our Lassie . .	F.	308	2	2	Lassie, 101 = 2.
Our Queen . .	M.	287	8	A.	Queen, 80 = 8.
Out o' Sight . .	C.	874	1	3	
Outpass . . .	G.	548	8	5	
Outpost . . .	H.	953	8	A.	
Outsider . . .	G.	671	5	3	
Overbury . . .	H.	499	4	5	
Over Norton . .	H.	989	8	5	Over, 287 = 8.

Name.	Sex.	Value.	Remainder.	Age (figures indicate years and "A" aged).	Remarks.
Overrated . .	M.	901	1	A.	
Overture . . .	M.	490	4	A.	
Owston Wood . .	C.	527	5	2	Owston, 517 = 4.
Oxhill . . .	G.	117	9	A.	
Oxonian . . .	H.	228	3	5	
Paddie Moore .	F.	341	8	3	Paddie, 95 = 5.
Padilla . . .	F.	116	8	2	
Padishah . . .	G.	391	4	A.	
Padlock II. . .	C.	137	2	4	
Padlock III. . .	M.	137	2	6	
Paducah . . .	F.	116	8	3	
Pagan . . .	C.	151	7	2	
Page . . .	C.	93	3	2	
Pageant . . .	C.	534	3	2	
Paglia . . .	C.	142	7	2	
Pain-bis . . .	C.	152	8	2	
Paiute . . .	C.	497	2	4	
Palinurus . . .	G.	427	4	A.	
Palmaro . . .	H.	350	7	5	
Palm Beach . .	F.	136	1	2	Beach, 15 = 6.
Palmerston . .	G.	831	3	6	
Palmira . . .	M.	362	2	6	
Palmitine . .	F.	611	8	2	
Palmleaf (By) . .	H.	241	7	6	
Palm Lily . .	F.	191	2	4	Lily, 70 = 7.
Palm Sunday II. .	C.	245	2	4	Sunday, 124 = 7.
Pangloss . . .	C.	243	9	4	
Pantheon . . .	H.	596	2	5	
Panther . . .	G.	736	7	A.	
Pants . . .	H.	591	6	6	
Panzerona . .	M.	405	9	A.	
Papdale . . .	M.	205	7	5	
Papola . . .	M.	198	9	6	
Paradejunker . .	G.	567	9	5	
Paragon . . .	G.	352	1	4	
Parakeet . . .	M.	712	1	5	
Paraphrase . .	F.	579	3	2	
Parcel . . .	M.	371	2	6	
Pardon Monsieur .	C.	651	3	4	Pardon, 341 = 8.
Paregoric . . .	C.	537	6	2	
Parma Violet . .	M.	858	3	A.	Parma, 322 = 7.
Parody . . .	F.	297	4	3	
Parquetry . .	C.	921	3	5	

Name.	Sex.	Value.	Remainder.	Age (figures indicate years and "A" aged).	Remarks.
Paramatta	H.	724	4	A.	
Parrot	C.	681	6	3	
Parsival	C.	451	1	4	
Parthian II.	H.	746	8	6	
Partridge	G.	884	2	4	
Partridge Green	G.	1164	3	5	Green, 280=1.
Passaro	C.	347	5	4	
Pas Seul (By)	G.	241	7	6	
Passion Flower	G.	749	2	A.	Flower, 318=3.
Past Master	H.	1242	9	6	Master, 701=8.
Pat-a-Cake	G.	532	1	A.	
Patlander	G.	766	1	6	
Patrick's Ball	G.	795	3	6	Ball, 34=7.
Patron (By)	F.	731	2	2	
Patron Saint	C.	1251	9	4	Saint, 520=7.
Patton	G.	531	9	6	
Paul II.	G.	112	4	4	
Paul Kendal	H.	226	1	A.	Paul, 112=4.
Pauloff	G.	198	9	5	
Paul-Puk-Keewis	C.	406	1	2	
Pavement	C.	596	2	4	
Pavillon	C.	251	8	4	
Pavo	G.	167	5	A.	
Pawnbroker	G.	560	2	A.	
Pax	H.	191	2	6	
Paxton	G.	641	2	2	
Paysanne	M.	201	3	6	
Peace and Plenty	G.	785	2	A.	Peace, 150=6.
Pearl Pin	M.	440	8	5	Pin, 130=4.
Pearl Rover	H.	796	4	A.	Rover, 486=9.
Pearly Reid	F.	534	3	4	Reid, 214=7.
Pecadillo	G.	151	7	6	
Peccavi	G.	201	3	A.	
Peculator	H.	757	1	5	
Pedlar	G.	324	9	A.	
Pedometer	M.	750	3	5	
Peep O II.	G.	177	6	A.	Peep, 170=8.
Peer Gynt	C.	743	5	2	Peer, 290=2.
Pekin	C.	160	7	3	
Pelargonium	F.	447	6	4	
Pellisson	H.	232	7	5	
Penant	F.	591	6	3	
Penarth	M.	746	8	5	
Pendulum	F.	214	7	3	

Name.	Sex.	Value.	Remainder.	Age (figures indicate years and "A" aged).	Remarks.
Penguin . . .	F.	216	9	3	
Peninsula . .	G.	281	2	A.	
Penitence . .	F.	660	3	4	
Pentland . .	G.	625	4	5	
Penwiper . .	F.	426	3	2	
Peopleton . .	G.	650	2	A.	
Pepita II.. .	M.	571	4	A.	
Pepper and Salt (By)	M.	839	2	A.	Pepper, 292=4.
Pepsaline . .	G.	320	5	6	
Perdicus . .	G.	364	4	5	
Perfectionist .	C.	1200	3	3	
Perfidious .	M.	434	2	5	
Pericles . .	C.	357	6	3	
Peridane . .	F.	354	3	4	
Perigueux . .	C.	333	9	4	
Period . .	G.	304	7	3	
Peripatetic .	G.	1201	4	A.	
Peripied . .	F.	384	6	3	
Periwinkle . .	F.	396	9	2	
Perseus . .	C.	410	5	3	
Perseus II. .	G.	410	5	2	
Perseverance .	G.	741	3	5	
Persiflage .	G.	463	4	3	
Persimmon (By) .	C.	430	7	2	
——— . .	F.	430	7	2	
Personality .	F.	831	3	2	
Perth Lad . .	G.	720	9	A.	Lad, 35=8.
Peruke . .	G.	316	1	4	
Petar . .		691	7	2	
Peter . .	G.	690	6	A.	
Peter II.. .	G.	690	6	A.	
Peter III. .	G.	690	6	5	
Peterina . .	M.	751	4	5	
Petersfield II..	G.	874	1	A.	
Petridge . .	H.	693	9	5	
Petriolo . .	G.	762	6	5	
Petronius. .	C.	816	6	4	
Petruchio II. .	G.	732	3	5	
Petunia . .	F.	757	1	3	
Phantom Knight .	G.	1021	4	4	Knight, 450=9.
Pharisee . .	C.	351	9	3	
Philip II.. .	G.	190	1	A.	
Philip III. .	G.	190	1	3	
Philsmead . .	G.	224	8	3	

Name.	Sex.	Value.	Remainder.	Age (figures indicate years and "A" aged).	Remarks.
Philter . . .	C.	710	8	2	
Phlegethon . .	G.	588	3	4	
Phosphor . . .	G.	422	8	A.	
Photius . . .	H.	446	5	5	
Phylloxera . .	G.	433	1	3	
Piano . . .	F.	147	3	2	
Picador . . .	C.	311	5	4	
Pickles . . .	H.	190	1	5	
Pickles . . .	G.	190	1	A.	
Picotee . . .	M.	516	3	A.	
Picotee . . .	M.	516	3	A.	
Pierre . . .	G.	290	2	4	
Pietermaritzburg .	C.	1560	3	4	
Pietra Santa . .	F.	1203	6	4	Pietra, 691=7.
Pikeman . . .	G.	200	2	5	
Pimlico . . .	G.	176	5	5	
Pimpernel . .	F.	490	4	3	
Pin	C.	130	4	2	
Pindar . . .	H.	335	2	6	
Pinefinch . . .	C.	263	2	4	
Ping Pong . .	C.	302	5	4	
Pinvin . . .	C.	260	8	3	
Pinzon (By) . .	F.	189	9	2	
Pioneer Queen .	F.	421	7	3	Queen, 80=8.
Pirate's Bride .	F.	903	3	4	Bride, 206=8.
Pirate Song . .	C.	822	3	2	Song, 132=6.
Pisa	M.	141	6	6	
Pistol . . .	C.	570	3	3	
Pit a Pat . . .	C.	962	8	2	
Pitch Dark . .	F.	308	2	2	
Pizzicato . . .	M.	133	7	5	
Plain Nan . .	M.	271	1	5	Nan, 101=2.
Platelayer . .	G.	760	4	4	
Plato . . .	G.	126	9	6	
Play-time . . .	G.	560	2	6	
Pleasure II. . .	G.	327	3	6	
Pledge . . .	F.	123	6	4	
Plenty of Time .	M.	1101	3	A.	Plenty, 580=4.
Plight . . .	F.	510	6	2	
Plumage . . .	F.	160	7	4	
Plum Pecker . .	C.	460	1	2	Pecker, 310=4.
Plunger . . .	G.	363	3	5	
Pluto . . .	G.	131	5	5	
Poetaster . . .	G.	1157	5	A.	

Name.	Sex.	Value.	Remainder.	Age (figures indicate years and "A" aged).	Remarks.
Poetess . . .	F.	556	7	2	
Point du Jour . .	C.	354	3	3	
Polak . . .	C.	137	2	3	
Pole Carew . .	C.	342	9	3	
Polestar . . .	F.	786	3	3	
Polin . . .	G.	166	4	3	
Poll Tax . . .	H.	593	8	A.	
Pomfret . . .	G.	812	2	5	
Pompeii . . .	F.	223	7	3	
Pom Pom . .	C.	244	1	3	
Pompous . . .	H.	262	1	A.	
Poor Pat . . .	G.	767	2	5	Pat, 481 = 4.
Poor Thing . .	F.	761	5	2	Thing, 475 = 7.
Poplar Grove . .	F.	698	5	3	Grove, 306 = 9.
Poppito . . .	F.	544	4	3	
Porcelaine . .	G.	436	4	3	
Pork Pie . .	C.	396	9	4	Pie, 90 = 9.
Porphyrion . .	G.	622	1	6	
Port Blair . .	C.	928	1	3	Port, 686 = 2.
Portcullis . . .	C.	796	4	4	
Port Erin . . .	M.	947	2	5	Port, 686 = 2.
Port Jackson . .	F.	820	1	2	Port, 686 = 2.
Portora . . .	G.	893	2	5	
Portroyal . . .	H.	928	1	A.	
Porus . . .	C.	346	4	4	
Posilipo . . .	G.	258	6	4	
Possible . . .	G.	174	3	A.	
Postman's Knock .	C.	768	3	4	Knock, 72 = 9.
Post Obit . .	C.	1150	7	2	Obit, 404 = 8.
Post Restante (By) .	C.	1267	7	2	
Potboy . . .	G.	496	1	A.	
Potch . . .	G.	85	4	4	
Potin . . .	C.	436	4	4	
Powick . . .	G.	108	9	A.	
Pradella . . .	F.	326	2	3	
Prairie Flower . .	F.	818	8	2	Flower, 318 = 3.
Prairie Rose . .	M.	713	2	6	
Prairie Guide . .	M.	524	2	6	Guide, 24 = 6.
Precocious . .	C.	672	6	4	
Preen . . .	C.	340	7	3	
Premature . .	G.	534	3	6	
Premier II. . .	G.	540	9	6	
Presbyterian . .	C.	1032	6	4	
Prescraggan . .	G.	641	2	2	

Name.	Sex.	Value.	Remainder.	Age (figures indicate years and "A" aged).	Remarks.
Presgrave . .	G.	660	3	5	
President . .	H.	814	4	A.	
President Roosevelt	C.	1600	7	2	Roosevelt, 786 = 3.
President Steyn .	C.	1334	2	4	Steyn, 520 = 7.
Preston Gate . .	F.	1230	6	2	Gate, 430 = 7.
Presumptive . .	C.	817	7	2	
Pretender . .	G.	954	9	6	
Pretty Alice . .	F.	410	5	4	Alice, 101 = 2.
Pretty Fair . .	F.	599	5	4	Fair, 290 = 2.
Pretty Garter . .	F.	1130	5	4	Garter, 821 = 2.
Pride (By) . .	C.	284	5	2	
——— . .	F.	284	5	2	
——— . .	C.	284	5	2	
Pride of Mabestown	M.	924	6	6	Pride, 284 = 5.
Pride of Windermere	F.	875	2	3	Pride, 284 = 5.
Priestess II. . .	M.	820	1	A.	
Priestlaw . .	G.	781	7	6	
Prime Alice . .	F.	421	7	4	Alice, 101 = 2.
Primrose II. .	M.	533	2	6	
Primrose Maid .	M.	587	2	6	Maid, 54 = 9.
Prince . . .	G.	390	3	A.	
Prince Chalcis .	C.	544	4	4	Chalcis, 154 = 1.
Prince Charming .	H.	704	2	5	Charming, 314 = 8.
Prince Florizel .	C.	753	6	3	Florizel, 363 = 3.
Prince George .	C.	598	4	4	George, 208 = 1.
Prince Hampton II.	H.	886	4	5	Hampton, 496 = 1.
Prince Imperial .	H.	761	5	6	Prince, 390 = 3.
Prince Leo .	C.	436	4	4	Leo, 46 = 1.
Prince Llewellyn .	G.	516	3	4	Llewellyn, 126 = 9.
Prince Melton .	C.	920	2	4	Melton, 530 = 8.
Prince of Monaco .	C.	594	9	2	Prince, 390 = 3.
Prince of Naples .	H.	701	8	6	Prince, 390 = 3.
Prince Regent .	C.	1053	9	4	Prince, 390 = 3.
Princesimmon .	C.	550	1	3	
Prince Talleyrand .	C.	1086	6	4	Prince, 390 = 3.
Prince Tuscan .	G.	920	2	A.	Tuscan, 530 = 8.
Princess . . .	F.	460	1	4	
Princess II. .	M.	460	1	5	
Princess Amie .	F.	521	8	4	Amie, 61 = 7.
Princess Hampton .	F.	956	2	3	Hampton, 496 = 1.
Princess Hilda .	M.	500	5	A.	Hilda, 40 = 4.
Princess May III. .	M.	510	6	A.	May, 50 = 5.
Princess Melton	F.	990	9	4	Melton, 530 = 8.
Princess of Ayr .	F.	752	5	3	Princess, 460 = 1.

Name.	Sex.	Value.	Remainder.	Age (figures indicate years and "A" aged).	Remarks.
Princess Olga . .	M.	518	5	6	Olga, 58=4.
Princess Ottilia .	F.	908	8	3	Ottilia, 448=7.
Princess Rhue . .	F.	666	9	4	Rhue, 206=8.
Princess Sophie .	F.	616	4	2	Sophie, 156=3.
Princess Teck . .	M.	890	8	5	Teck, 430=7.
Principality . .	F.	911	2	2	
Prior's Fancy . .	F.	689	5	3	Fancy, 201=3.
Prisoner (By) . .	C.	537	6	3	
—— . .	C.	537	6	2	
—— . .	F.	537	6	3	
Prisoner of Zenda .	C.	690	6	2	Prisoner, 537=6.
Privado . . .	H.	371	2	A.	
Prize Cherry . .	F.	510	6	4	Cherry, 223=7.
Professor II. . .	G.	636	6	5	
Prohibition . .	G.	643	4	5	
Proofsheet . .	F.	1076	5	4	
Prorogation . .	G.	863	8	4	
Prose . . .	H.	293	5	A.	
Prosper . . .	M.	622	1	A.	
Prosset . . .	G.	752	5	A.	
Protest . . .	C.	1156	4	3	
Proteus . . .	C.	756	9	2	
Proud Agnes . .	M.	431	8	5	Agnes, 141=6.
Proud Chieftain .	H.	833	5	A.	Chieftain, 543=3.
Proudfute . .	C.	776	2	3	
Proud Star . .	C.	951	6	4	Star, 661=4.
Proxime . . .	F.	442	1	3	
Prudence . . .	M.	400	4	6	
Puerto . . .	G.	311	5	4	
Pull-away . .	C.	127	1	3	
Pumpernickel (By) .	F.	500	5	3	
Puna . . .	G.	137	2	4	
Punch Ladle . .	G.	198	9	A.	Ladle, 65=2.
Punctilio . .	F.	596	2	3	
Pure Gold . .	C.	355	5	3	
Pure Joy . .	G.	311	5	3	
Purse . . .	G.	340	7	A.	
Pursuit . . .	F.	746	8	2	
Pyperstone . .	G.	876	3	4	
Pyramid II. . .	M.	325	1	A.	
Pyswell . . .	G.	186	6	5	
Quadrant . .	C.	681	6	2	
Quadruped . .	G.	325	1	A.	

Name.	Sex.	Value.	Remainder.	Age (figures indicate years and "A" aged).	Remarks.
Quaich . . .	F.	330	6	3	
Quality Tells . .	G.	967	4	A.	
Quartz . . .	M.	634	4	5	
Queen Audacia .	F.	158	5	3	Queen, 80=8.
Queen Catherine .	F.	766	1	4	Catherine, 686=2.
Queengold . .	F.	140	5	4	
Queen Joan . .	F.	139	4	4	Joan, 59=5.
Queen of Coins .	F.	293	5	3	Queen, 80=8.
Queen of the Gipsies	M.	345	3	6	Queen, 80=8.
Queen of the Moor .	F.	421	7	4	Queen, 80=8.
Queen Regent . .	M.	743	5	A.	Queen, 80=8.
Queen's Birthday (By) . . .	G.	708	6	4	Birthday, 621=9.
Queen's Bower .	F.	297	9	3	Bower, 210=3.
Queen's Copper .	F.	311	5	4	Copper, 224=8.
Queen's Counsel (By)	F.	253	1	3	Counsel, 166=4.
Queen's Gold . .	G.	147	3	5	Gold, 60=6.
Queen's Key . .	G.	117	9	2	Key, 30=3.
Queen's Park . .	G.	388	1	A.	Park, 301=4.
Queen Theo . .	F.	501	6	4	Theo, 421=7.
Query . . .	G.	240	6	A.	
Quichart . . .	G.	624	3	6	
Quick Change . .	M.	97	7	5	
Quickly Wise (By) .	F.	93	3	3	Wise, 13=4.
—— . . .	F.	93	3	2	Wise, 13=4.
Quick Shot . .	G.	742	4	5	
Quicksight . .	F.	500	5	4	
Quicksilver II.. .	M.	410	5	A.	
Quickstep . .	G.	590	5	A.	
Quick Wit . .	M.	446	5	5	
Quiff . . .	F.	100	1	4	
Quintessence . .	F.	660	3	2	
Quintus Fabius .	C.	683	8	3	
Quirites . . .	G.	990	6	5	
Raasay . . .	C.	272	2	4	
Rabelais . . .	C.	253	1	2	
Racoon . . .	M.	277	7	5	
Racine . . .	C.	321	6	3	
Rack a Rock . .	C.	444	3	3	Rack, 221=5.
Racket . . .	G.	631	1	5	
Radcliffe . . .	G.	335	2	A.	
Radegonde II.. .	F.	291	3	4	
Radical II. . .	G.	255	3	5	

Name.	Sex.	Value.	Remainder.	Age (figures indicate years and "A" aged).	Remarks.
Radieuse II. . .	F.	228	3	3	
Radivia . . .	F.	296	8	3	
Radius (By) . .	F.	275	5	2	
Radmore . . .	G.	451	1	A.	
Radnage . . .	F.	258	6	3	
Raeburn (By) . .	F.	462	3	2	
—— . . .	F.	462	3	2	
—— . . .	G.	462	3	2	
—— . . .	C.	462	3	2	
—— . . .	F.	462	3	3	
—— . . .	F.	462	3	2	
—— . . .	G.	462	3	3	
Rafale . . .	F.	321	6	4	
Raferagh . . .	F.	492	6	2	
Raft	H.	681	6	5	
Ragamuffin . .	C.	392	5	3	
Raggy . . .	F.	231	6	4	
Ragimunde (By) .	F.	308	2	2	
Ragwort . . .	F.	827	8	3	
Raiden . . .	G.	264	3	3	
Rail	G.	240	6	A.	
Rain Gauge . .	C.	293	5	2	
Rainstorm . .	C.	962	8	4	
Rainton . . .	G.	710	8	A.	
Raise . . .	M.	217	1	5	
Rally . . .	H.	241	7	6	
Rambling Katie .	M.	774	9	5	Katie, 431 = 8.
Rambling Mary .	M.	594	9	6	Mary, 251 = 8.
Ramondia . .	F.	308	2	3	
Ram Sing . .	C.	371	2	4	
Rancher . . .	C.	454	4	4	
Randle . . .	G.	285	6	6	
Ranger II. . .	G.	454	4	A.	
Ranunculus . .	G.	411	6	4	
Raphallo (By) . .	C.	317	2	2	
Rapide . . .	C.	295	7	2	
Rapparee . . .	H.	413	8	5	
Rappel . . .	F.	243	9	2	
Rapture . . .	M.	484	7	6	
Ras Makunnen .	C.	428	5	4	Ras, 261 = 9.
Rathburne . .	C.	858	3	3	
Rathcline . . .	G.	716	5	A.	
Rathcroggan . .	C.	902	2	2	
Rather Warm . .	G.	652	4	2	Warm, 247 = 4.

Name.			Sex.	Value.	Remainder.	Age (figures indicate years and "A" aged).	Remarks.
Rathgowan	.	.	G.	682	7	6	
Rathkeale	.	.	G.	666	9	A.	
Rathmaeve	.	.	G.	736	7	A.	
Rathmoyle	.	.	C.	688	4	2	
Rathmullen	.	.	G.	726	6	A.	
Rattler II.	.	.	G.	831	3	A.	
Ravager	.	.	G.	494	8	5	
Ravarnet	.	.	F.	942	6	4	
Ravel	.	.	C.	311	5	2	
Raveno	.	.	C.	347	5	3	
Ravenous	.	.	G.	401	5	3	
Ravensbury (By)		.	F.	613	1	2	
———	.	.	G.	613	1	2	
———	.	.	F.	613	1	2	
———	.	.	F.	613	1	2	
———	.	.	C.	613	1	2	
———	.	.	F.	613	1	2	
Ravenscar	.	.	C.	622	1	3	
Ravensclaw	.	.	C.	453	3	3	
Ravenscliff	.	.	G.	531	9	4	
Ravensdale II.	.	.	G.	445	4	6	
Raven's Flight	.	.	C.	911	2	3	
Ravensheugh	.	.	H.	442	8	5	
Ravenside	.	.	G.	405	9	5	
Ravensroost	.	.	F.	1067	5	3	
Ravina	.	.	F.	332	8	3	
Ravola	.	.	M.	318	3	5	
Ray	.	.	C.	210	3	2	
Rayleigh	.	.	C.	250	7	3	
Raymond	.	.	G.	304	7	6	
Ray's Cross	.	.	C.	489	4	4	Cross, 282 = 3.
Raysos	.	.	G.	332	8	6	
Ready	.	.	F.	224	8	4	
Rebecca	.	.	F.	233	8	4	
Rebecca II.	.	.	M.	233	8	5	
Rebecca III.	.	.	M.	233	8	A.	
Rebel Boy	.	.	C.	256	4	4	Boy, 14 = 5.
Recanter	.	.	C.	881	8	4	
Reckitt	.	.	G.	630	9	2	
Red Axe	.	.	H.	335	2	5	Axe, 121 = 4.
Red B.	.	.	M.	226	1	A.	B, 12 = 3.
Red Bell	.	.	F.	256	4	4	Bell, 42 = 6.
Red Cedar	.	.	C.	489	3	3	Cedar, 275 = 5.
Red Cloak	.	.	M.	290	2	A.	Cloak, 76 = 4.

Name.	Sex.	Value.	Remainder.	Age (figures indicate years and " A " aged).	Remarks.
Red Comyn . .	C.	326	2	3	Comyn, 112=4.
Red Duke . .	C.	254	2	2	Duke, 40=4.
Red Eager . .	C.	445	4	2	Eager, 231=6.
Red Friar II. . .	G.	695	2	A.	Friar, 481=4.
Red Hall . .	G.	251	8	5	Hall, 37=1.
Red Hot . .	G.	621	9	A.	Hot, 407=2.
Red Letter .	G.	854	8	A.	Letter, 640=1.
Red Light . .	H.	644	5	5	Light, 430=7.
Red Lily . .	F.	284	5	2	Lily, 70=7.
Red Ocean .	F.	581	5	4	Ocean, 367=7.
Red Prince II. .	F.	604	1	3	Prince, 390=3.
Red Rag . .	M.	435	3	6	Rag, 221=5.
Red Reel . .	G.	454	4	5	Reel, 240=6.
Red Rice . .	F.	474	6	2	Rice, 260=8.
Red Robber .	H.	618	6	5	Robber, 404=8.
Red Robber II. .	G.	618	6	A.	Robber, 404=8.
Redskin II. . .	G.	344	2	A.	
Red, White and Blue II. . .	G.	713	2	A.	Red, 214=7.
Red Wing . .	M.	290	2	5	Wing, 76=4.
Reeve . .	H.	290	2	A.	
Refinery . .	F.	550	1	3	
Regalia . .	F.	272	2	3	
Regent . .	H.	633	6	6	
Rehearsal. .	C.	505	1	3	
Reine des Fleurs .	F.	594	9	4	Reine, 260=8.
Reine Margot .	F.	527	5	2	Reine, 260=8.
Reitz . .	H.	617	5	5	
Rejected Memory .	M.	947	2	A.	Memory, 300=3.
Religion II. . .	F.	293	5	4	
Renzo . .	C.	273	3	3	
Repel . .	C.	330	6	2	
Repentress .	F.	1020	3	4	
Reredos . .	C.	486	9	A.	
Research . .	G.	473	5	5	
Residant . .	C.	734	5	4	
Resident Magistrate	G.	1847	2	A.	Magistrate, 1113=6.
Resistance .	F.	787	4	3	
Restless . .	F.	770	5	3	
Restored . .	G.	880	7	A.	
Retz . .	C.	617	5	3	
Reuben . .	G.	278	8	5	
Reuben II. .	G.	278	8	6	
Reveller . .	G.	530	8	4	

Name.	Sex.	Value.	Remainder.	Age (figures indicate years and "A" aged).	Remarks.
Revelry . . .	C.	530	8	3	
Revenue . . .	C.	356	5	4	
Revera . . .	F.	501	6	4	
Reverie II. . .	M.	510	6	6	
Reverse II. . .	M.	550	1	5	
Reversed . . .	G.	554	5	4	
Revival . . .	M.	400	4	6	
Revoke . . .	H.	316	1	5	
Revolution . .	F.	692	8	4	
Revolutionaire . .	C.	902	2	3	
Revolver . . .	F.	602	8	4	
Rhine Violet . .	M.	796	4	6	
Rhodesia . . .	M.	291	3	A.	
Rhomboid . .	H.	250	7	5	
Rhyall Lass . .		333	9	4	Lass, 91 = 1.
Rhyton . . .	G.	660	3	5	
Riccarton . .	H.	871	7	5	
Rice	H.	260	8	5	
Rickenhore . .	G.	481	4	A.	
Rickman . . .	C.	310	4	4	
Ridiculous . .	F.	314	8	3	
Riding Master . .	C.	975	3	3	
Rifleman II. . .	G.	401	5	A.	
Right Away (By) .	G.	617	5	2	
—— . . .	C.	617	5	3	
—— . . .	F.	617	5	2	
—— . . .	F.	617	5	2	
Rightful . . .	C.	710	8	3	
Right of Way . .	C.	697	4	2	
Rigo	G.	226	1	4	
Rigolet . . .	H.	266	5	6	
Ringburn . . .	C.	522	9	2	
Ringdrake . .	C.	504	9	3	
Ringing the Changes	G.	418	4	3	
Ringleader . .	G.	514	1	5	
Ringville . . .	G.	390	3	5	
Rio Grand . .	C.	487	1	3	Rio, 216 = 9.
Ripe and Ready .	C.	559	1	2	Ripe, 280 = 1.
Risboro . . .	C.	470	2	3	
Risby . . .	M.	272	2	5	
Risette . . .	F.	670	4	3	
Rising Falcon . .	F.	458	8	2	Falcon, 181 = 1.
Rising Glass . .	C.	388	1	3	Glass, 111 = 3.
River Side II. .	C.	544	4	2	

Name.	Sex.	Value.	Remainder.	Age (figures indicate years and "A," aged).	Remarks.
Robert le Diable	C.	505	1	3	Robert, 418 = 4.
Robin Hood	G.	263	2	4	
Robin Hood II.	G.	263	2	6	
Robino	G.	268	7	5	
Robin Roan	G.	510	6	5	Robin, 254 = 2.
Rob Roy	C.	416	2	3	
Rob Roy	C.	416	2	4	
Rob Roy III.	G.	416	2	A.	
Rochdale	G.	266	5	A.	
Rochelle	M.	546	6	5	
Rockabill	G.	255	3	4	
Rock Castle	G.	333	9	4	Castle, 111 = 3.
Rocket II.	G.	632	2	A.	
Rockferry	H.	522	9	5	
Rockhill	F.	257	5	4	
Rock Sand	C.	337	4	2	Sand, 115 = 7.
Rockview	G.	318	3	5	
Rococo	G.	254	2	A.	
Roderic	G.	426	3	6	
Rodomel	G.	292	4	A.	
Roe O'Neill	C.	303	6	2	Roe, 206 = 8.
Rohotaranga	H.	889	7	A.	
Roman Child	G.	333	9	5	Child, 37 = 1.
Romancer II.	G.	557	8	4	
Roman Empress	F.	697	4	4	Empress, 401 = 5.
Romanoff	G.	378	9	A.	
Roman Queen	F.	376	7	2	Queen, 80 = 8.
Romeo		262	1	5	
Ronald	G.	286	7	2	
Rontgen	H.	305	8	A.	
Roover Crag	C.	727	7	2	
Rosario	H.	430	7	6	
Roseberry	M.	435	3	A.	
Rose Blair	F.	455	5	3	Rose, 213 = 6.
Roseburn	C.	465	6	2	
Rose Coon	F.	289	1	2	
Rosedale	C.	257	5	2	
Rosegarland II.	M.	518	5	A.	
Rose Gift	M.	713	2	5	Rose, 213 = 6.
Rose Graft		914	5	5	Rose, 213 = 6.
Rose Kite	M.	633	3	4	
Rose of England	M.	449	8	5	Rose, 213 = 6.
Roseville	G.	333	9	A.	
Rosey O'More	M.	476	8	A.	Rosey, 223 = 7.

Name.	Sex.	Value.	Remainder.	Age (figures indicate years and "A" aged).	Remarks.
Rose Window (By) .	C.	279	9	2	Window, 66=3.
—— . . .	F.	279	9	2	Window, 66=3.
—— . . .	F.	279	9	2	Window, 66=3.
Rose Wreath . .	G.	828	9	6	Wreath, 615=3.
Rosezeta . . .	M.	641	2	5	
Rosglas . . .	H.	373	4	6	
Roshven . . .	H.	642	3	5	
Rosie Doon . .	F.	283	4	3	Rosie, 223=7.
Rosie O'Grady. .	F.	465	6	3	Rosie, 223=7.
Rosina . . .	M.	264	3	A.	
Roskey Lad . .	G.	327	3	A.	
Rossall . . .	G.	292	4	3	
Rostrevor . .	H.	1152	9	5	
Roswall (By) . .	C.	308	2	2	
Rotten Row . .	G.	858	3	2	Rotten, 652=4.
Rouge . . .	M.	506	2	6	
Rough and Ready (By)	C.	559	1	2	Rough, 280=1.
Roughborough .	H.	490	4	A.	
Roughside . .	H.	344	2	A.	
Round Robin . .	G.	514	1	4	Robin, 254=2.
Rouster . . .	G.	866	2	4	
Row	M.	206	8	6	
Rowfant . . .	C.	737	8	2	
Roy	G.	212	5	5	
Royal Bevy . .	C.	344	2	2	Bevy, 102=3.
Royal Betty . .	F.	273	3	4	Betty, 31=4.
Royal Britain . .	H.	894	3	5	Britain, 652=4.
Royal Chieftain .	G.	785	2	3	Chieftain, 543=3.
Royal Crescent .	C.	982	1	4	Crescent, 740=2.
Royal Cygnet . .	C.	772	7	2	Cygnet, 530=8.
Royal Dance . .	G.	357	6	5	Dance, 115=7.
Royal Dane . .	C.	306	9	3	Dane, 64=1.
Royal Devon . .	F.	386	8	4	Devon, 144=9.
Royal Don . .	G.	298	1	3	
Royal Drake . .	G.	476	8	4	Drake, 234=9.
Royal Emperor II. .	M.	773	8	A.	Emperor, 531=9.
Royal Esher . .	G.	753	6	3	Esher, 511=7.
Royal George . .	C.	450	9	4	George, 208=1.
Royal George . .	G.	450	9	5	George, 208=1.
Royal Holly . .	C.	289	1	3	Holly, 47=2.
Royal Joy . .	C.	257	5	3	Joy, 15=6.
Royal Lancer . .	C.	385	7	3	Lancer, 142=7.
Royal Lineage . .	H.	335	2	6	Lineage, 93=3.

Name.	Sex.	Value.	Remainder.	Age (figures indicate years and "A" aged).	Remarks.
Royal Mantle . .	F.	763	7	4	Mantle, 521=8.
Royal Minster . .	C.	992	2	3	Minster, 750=3.
Royal Monk . .	G.	354	3	2	Monk, 112=4.
Royal Offspring .	G.	734	5	6	Offspring, 492=6.
Royal Palm . .		363	3	6	Palm, 121=4.
Royal Plume . .	C.	398	2	4	Plume, 156=3.
Royal Quiver . .	G.	542	2	A.	Quiver, 300=3.
Royal Rays . .	H.	459	9	5	Rays, 217=1.
Royal River . .	C.	722	2	4	River, 480=3.
Royal Rouge . .	C.	748	1	4	Rouge, 506=2.
Royal Serf . .	C.	582	6	2	Serf, 340=7.
Royal Songster .	G.	1034	8	6	Songster, 792=9.
Royal Sovereign (By)	F.	644	5	2	Sovereign, 402=6.
Royal Stream . .	C.	952	7	3	Stream, 710=8.
Royal Suiter . .	G.	908	8	4	Suiter, 666=9.
Royal Summons .	F.	452	2	4	Summons, 210=3.
Royalty II. . .	M.	652	4	A.	
Royalty III. . .	M.	652	4	A.	
Royal Warden . .	H.	503	8	A.	Warden, 261=9.
Royal Winkfield .	C.	442	1	4	Winkfield, 200=2.
Royette . . .	M.	612	9	A.	
Royston Glen . .	G.	832	4	4	Royston, 722=2.
Royston Prince .	H.	1112	5	5	Royston, 722=2.
Rubbish . . .	G.	502	7	2	
Rubiana . . .	F.	264	3	2	
Ruby Ray . .	G.	428	5	4	Ruby, 218=2.
Rufus II. . .	G.	346	4	A.	
Rummager . .	G.	444	3	A.	
Runaway Girl . .	M.	517	4	5	Girl, 250=7.
Running Stream .	F.	1030	4	4	Stream, 710=8.
Ruscombe . .	G.	326	2	5	
Rush (By) . .	G.	500	5	2	
—— . . .	G.	500	5	2	
—— . . .	F.	500	5	2	
Rushgate . . .	G.	931	4	2	
Rushlight . . .	F.	930	3	2	
Rushport . . .	C.	1186	7	2	
Rushyford . . .	C.	794	2	2	
Ruskin . . .	C.	330	6	4	
Russet Brown . .	M.	918	9	6	
Rutland . . .	G.	685	1	5	
Ruy Blas III. . .	C.	243	9	3	
Ruy Lopez . .	C.	343	1	2	
Ruyter . . .	C.	810	9	3	

Name.	Sex.	Value.	Remainder.	Age (figures indicate years and " A " aged).	Remarks.
Rye . . .	G.	210	3	A.	
Ryevale . .	G.	256	4	A.	
Ryhall Lass .		338	5	4	Lass, 91 = 1.
Sabbath . .	G.	462	3	5	
Sabot . .	C.	69	6	4	
Sabella . .	F.	104	5	3	
Sabine King .	G.	213	6	3	King, 90 = 9.
Sablewing .	M.	169	7	5	
Saccharine .	F.	341	8	4	
Saccharum .	M.	322	7	A.	
Sackcloth .	F.	538	7	3	
Sada . .	F.	66	3	2	
Saengrin . .	F.	337	4	4	
Sagaman . .	G.	172	1	5	
Sagittarius .	G.	735	6	A.	
Sailaway . .	G.	117	9	A.	
Sailor King .	G.	390	3	A.	King, 90 = 9.
St. Aldegonde .	F.	641	2	4	Aldegonde, 121 = 4.
St. Alwyne .	C.	617	5	3	Alwyne, 97 = 7.
St. Ambrose .	C.	776	2	3	Ambrose, 256.
St. Amour .	C.	767	2	4	Amour, 247 = 4.
St. Angelo (By) .	G.	620	8	2	Angelo, 100 = 1.
——— . .	F.	620	8	2	Angelo, 100 = 1.
——— . .	C.	620	8	4	Angelo, 100 = 1.
——— . .	C.	620	8	2	Angelo, 100 = 1.
——— . .	G.	620	8	3	Angelo, 100 = 1.
St. Antonius .	C.	706	4	3	Antonius, 186 = 6.
St. Asaph . .	C.	691	7	2	Asaph, 171 = 9.
St. Asaph the First .	G.	1245	3	A.	Asaph, 171 = 9.
St. Atholine .	F.	1022	5	2	Atholine, 502 = 7.
St. Aubyn .	G.	574	7	A.	
St. Barts . .	C.	723	3	2	
St. Benet . .	C.	992	2	3	Benet, 472 = 4.
St. Bernard II. .	G.	976	4	6	Bernard, 456 = 6.
St. Beurre .	H.	722	2	5	Beurre, 202 = 4.
St. Brendan .	C.	836	8	3	Brendan, 316 = 1.
St. Briavels .	C.	913	4	3	Briavels, 393 = 6.
St. Canice .	G.	651	3	5	Canice, 131 = 5.
St. Cassimir .	G.	841	4	5	Cassimir, 321 = 6.
St. Clears . .	G.	840	3	5	Clears, 320 = 5.
St. Colon . .	G.	626	5	4	Colon, 106 = 7.
St. Cuthbert .	H.	1547	8	6	Cuthbert, 1027 = 1.
Saintcraft . .	F.	1221	6	2	

Name.	Sex.	Value.	Remainder.	Age (figures indicate years and "A" aged).	Remarks.
St. David (By) . .	F.	535	4	3	David, 15=6.
——— . . .	G.	535	4	3	David, 15=6.
——— . . .	F.	535	4	3	David, 15=6.
St. Elias . . .	H.	622	1	6	Elias, 102=3.
St. Emilion . .	C.	661	4	2	Emilion, 141=6.
St. Erth . . .	G.	1126	1	5	Erth, 606=3.
St. Expedit . .	C.	1105	7	2	Expedit, 585=9.
St. Feveronia . .	F.	957	3	3	Feveronia, 437=5.
St. Florian (By) .	G.	896	5	3	Florian, 376=7.
St. Francis . .	H.	971	8	5	Francis, 451=1.
St. Fridolin . .	C.	890	8	3	Fridolin, 370=1.
St. Frusquin (By) .	F.	930	3	2	Frusquin, 410=5.
St. Gall . . .	H.	572	5	5	Gall, 52=7.
St. Galmier II.. .	G.	821	2	A.	Galmier, 301=4.
St. Gerald . .	C.	768	3	2	Gerald, 248=5.
St. Gothard . .	G.	750	3	5	Gothard, 230=5.
St. Hilaire (By) .	F.	775	1	2	Hilaire, 255=3.
——— . . .	F.	775	1	2	Hilaire, 255=3.
——— . . .	C.	775	1	2	Hilaire, 255=3.
——— . . .	F.	775	1	2	Hilaire, 255=3.
St. Hilarious . .	C.	826	7	3	Hilarious, 306=9.
St. Hubert . .	C.	1143	9	3	Hubert, 623=2.
St. Hylda . .	F.	560	2	2	Hilda, 40=4.
St. Jacques . .	H.	548	8	5	Jacques, 28=1.
St. Jessica . .	M.	614	2	A.	Jessica, 94=4.
St. John . . .	G.	575	8	4	John, 55=1.
St. Just II. . .	H.	983	2	A.	Just, 463=4.
St. Leger (By) .	F.	763	7	4	Leger, 243=9.
St. Leger . . .	G.	763	7	A.	Leger, 243=9.
St. Leonards . .	G.	874	1	A.	Leonards, 354=3.
St. Levan . . .	C.	690	6	4	Levan, 170=8.
St. Luke . . .	C.	576	9	3	Luke, 56=2.
St. Lunnaire . .	G.	810	9	4	Lunnaire, 290=2.
St. Maclou . .	C.	617	5	4	Maclou, 97=7.
St. Magdalen . .	F.	675	9	3	Magdalen, 155=2.
St. Malo . . .	C.	597	3	4	Malo, 77=5.
St. Maurice . .	C.	822	3	4	Maurice, 302=5.
St. Mick . . .	G.	580	4	6	Mick, 60=6.
St. Monans . .	C.	726	6	4	Monans, 206=8.
St. Moritz . .	G.	1169	8	6	Moritz, 649=1.
St. Nydia . . .	M.	585	9	5	Nydia, 65=2.
St. Patrick II.. .	G.	1221	6	A.	Patrick, 701=8.
St. Paulus . .	G.	692	8	5	Paulus, 172=1.
St. Prisca . . .	F.	881	8	3	Prisca, 361=1.

Name.	Sex.	Value.	Remainder.	Age (figures indicate years and "A" aged).	Remarks.
St. Quintin . . .	C.	1040	5	3	Quintin, 520=7.
St. Rosalie . . .	M.	774	9	6	Rosalie, 254=2.
St. Rose . . .	F.	733	4	3	Rose, 213=6.
St. Saen . . .	G.	641	2	A.	Saen, 121=4.
St. Saens . . .	C.	700	7	3	Saens, 180=9.
St. Salvador . .	G.	902	2	5	Salvador, 382=4.
St. Saulge . . .	C.	615	3	3	Saulge, 95=5.
St. Serf (By) . .	G.	860	5	3	Serf, 340=7.
—— . . .	C.	860	5	2	Serf, 340=7.
—— . . .	F.	860	5	2	Serf, 340=7.
St. Simon (By) .	F.	670	4	3	Simon, 150=6.
—— . . .	C.	670	4	2	Simon, 150=6.
—— . . .	C.	670	4	2	Simon, 150=6.
St. Tudno . .	G.	980	8	6	Tudno, 460=1.
St. Uncomber . .	F.	839	2	3	Uncomber, 319=4.
St. Valentine . .	G.	1141	7	3	Valentine, 621=9.
St. Vallier . .	C.	840	3	4	Vallier, 320=5.
St. Vincent . .	G.	1160	8	5	Vincent, 640=1.
St. Walshaw . .	C.	859	4	2	Walshaw, 339=6.
St. Windeline . .	F.	680	5	3	Windeline, 160=7.
Saintly Michael .	G.	660	3	A.	Michael, 100=1.
Saintly Mick . .	G.	620	8	A.	Mick, 60=6.
Saints' Conquest .	F.	1142	8	4	Conquest, 562=4.
Saintwell . . .	G.	566	8	6	
Sais	G.	121	4	6	
Sakuntala . .	F.	561	3	3	
Sale	M.	100	1	A.	
Saleratus . . .	C.	762	6	3	
Salina . . .	M.	142	7	5	
Sallypark . . .	C.	402	6	4	
Saltator II. . .	H.	1092	3	5	
Salt Tears . .	F.	1161	9	3	Tears, 670=4.
Salute . . .	G.	497	2	2	
Salvador . . .	H.	382	4	5	
Salvia . . .	F	182	2	4	
Sal Volatile . .	M.	648	9	5	Sal, 91=1.
Salzburg . . .	G.	320	5	4	
Sam	G.	101	2	A.	
Samaritan . .	C.	752	5	2	
Samiel . . .	F.	141	6	4	
Sanctissima II. .	M.	632	2	6	
Sandal Beat . .	G.	557	8	3	Sandal, 145=1.
Sandbag . . .	C.	138	3	4	
Sandboy . . .	C.	129	3	2	

Name.	Sex.	Value.	Remainder.	Age (figures indicate years and "A", aged).	Remarks.
Sandflake . .	F.	225	3	3	
Sandpiper II. . .	C.	475	7	4	
Sandy Bree . .	M.	337	4	5	Bree, 212 = 5.
Sangrado . . .	G.	342	9	A.	
Sankence . . .	F.	251	8	2	
San Jose . . .	C.	137	2	3	Jose, 26 = 8.
San Lucar . .	H.	362	2	A.	Lucar, 251 = 8.
San Paulo . .	G.	229	4	5	Paulo, 118 = 1.
Sans Atout . .	F.	978	6	4	Atout, 807 = 6.
Sans Escompte .	H.	704	2	5	Escompte, 533 = 2.
Sans Gene . .	H.	234	9	5	Gene, 63 = 9.
Sans Nom . .	C.	263	2	3	Nom, 92 = 2.
Sansome . . .	C.	211	4	3	
Santa Mimosa . .	F.	659	2	3	Mimosa, 147 = 3.
Santa Teresa . .	M.	1140	6	6	Teresa, 628 = **7**.
San Terenzo . .	C.	794	2	2	Terenzo, 683 = 8.
Santoi . . .	H.	523	1	5	
Sapper . . .	H.	263	2	A.	
Sapphira . . .	M.	342	9	5	
Sarah . . .	M.	262	1	A.	
Sarah II. . .	M.	262	1	6	
Sarah Gamp . .	F.	403	7	2	Sarah, 262 = 1.
Sardanapale . .	C.	437	5	3	
Sardus . . .	G.	325	1	A.	
Saroth . . .	F.	672	6	2	
Sarsenet . . .	M.	391	4	5	
Saskia . . .	F.	152	8	3	
Satanita . . .	F.	913	4	3	
Saturday II. . .	F.	675	9	3	
Satyr . . .	C.	670	4	3	
Saucy Jack . .	H.	156	3	5	Jack, 24 = 6.
Savilion . . .	C.	231	6	4	
Savone . . .	C.	197	8	3	
Savorno . . .	F.	403	7	4	
Sawdust . . .	F.	526	4	3	
Saxilby . . .	H.	183	3	5	
Saxon . . .	C.	191	2	4	
Scamp II. . .	G.	201	3	A.	
Scampanio . .	G.	258	6	A.	
Scappata . . .	F.	485	8	2	
Scarem . . .	F.	321	6	4	
Scarlet Runner II. .	F.	961	7	4	
Scars . . .	G.	341	8	6	
Scattergun . .	C.	751	4	2	

Name.	Sex.	Value.	Remainder.	Age (figures indicate years and "A" aged).	Remarks.
Sceptre . . .	F.	551	2	3	
Schechallion . .	G.	701	8	A.	
Schemer . . .	G.	650	2	A.	
Schiedam . . .	G.	355	4	A.	
School Bell . .	G.	158	5	6	Bell, 42 = 6.
Scotch Cap . .	C.	186	6	3	Cap, 101 = 2.
Scotch Cream . .	G.	355	4	3	Cream, 270 = 9.
Scotch-fir . .	M.	365	5	A.	Fir, 280 = 1.
Scotch Law . .	M.	117	9	6	Law, 32 = 5.
Scotchman II. .	G.	175	4	5	
Scotchman III.	G.	175	4	5	
Scotch Tweed .	C.	114	6	3	Tweed, 29 = 2.
Scott . . .	G.	482	5	A.	
Scottish Archer .	C.	1187	8	3	Archer, 405 = 9.
Scottish Lass .	F.	873	9	4	Lass, 91 = 1.
Scoundrel . .	G.	370	1	4	
Scullery Maid .	M.	394	7	A.	Maid, 54 = 9.
Scullion . . .	F.	170	8	4	
Seaflight . .	M.	580	4	5	
Sea Flower . .	M.	388	1	6	Flower, 318 = 3.
Sea Fog . . .	G.	172	7	A.	Fog, 102 = 3.
Seagull . . .	M.	120	3	A.	
Seahorse II. .	H.	337	4	6	
Sea Log . . .	C.	121	4	2	Log, 52 = 7.
Sea Lord . . .	G.	306	9	3	Lord, 236 = 2.
Sea Rose . . .	M.	283	4	5	Rose, 213 = 6.
Seaside . . .	G.	134	8	A.	
Seastorm . . .	F.	772	7	4	
Sea Thistle . .	F.	565	7	2	Thistle, 495 = 9.
Sea-wall . . .	G.	108	9	A.	Wall, 38 = 2.
Second Thoughts .	F.	1011	3	3	Thoughts, 867 = 3.
Secret Service .	H.	1100	2	A.	Service, 400 = 4.
Seer	G.	270	9	A.	
Segment . . .	G.	590	5	3	
Semi-colon . .	M.	226	1	5	
Semiramis . .	F.	411	6	4	
Sempronius (By) .	C.	516	3	2	
Senateur . . .	H.	561	3	6	
Senator . . .	H.	721	1	5	
Sennoris . . .	C.	386	8	3	
Sentinel II. . .	G.	610	8	A.	
Separation . .	F.	702	9	4	
Sequel II. . .	G.	130	4	A.	
Sergeant . . .	C.	714	3	3	

Name.	Sex.	Value.	Remainder.	Age (figures indicate years and " A " aged).	Remarks.
Seringapatam . . .	H.	852	6	6	
Sermon . . .	F.	350	8	2	
Servatrix . . .	F.	1021	4	3	
Servitor . . .	C.	940	4	4	
Servitude . . .	F.	750	3	4	
Servius . . .	G.	400	4	A.	
Servula . . .	F.	371	2	4	
Set Fair . . .	F.	760	4	2	
Seth . . .	G.	475	7	A.	
Severna . . .	F.	331	7	A.	
Shackleford . .	G.	635	5	A.	
Shaker . . .	G.	530	8	A.	
Shallon . . .	M.	381	3	A.	
Shamrock II. . .	G.	563	5	3	
Shangrotha II. (By)	C.	983	2	2	
—— .	F.	983	2	2	
Shangora . . .	F.	578	2	3	
Shannon Lass . .	M.	492	6	A.	Lass, 91 = 1.
Shan-si . . .	G.	421	7	5	
Sharcott . . .	C.	923	5	4	
Shaun Aboo . .	G.	365	5	4	Aboo, 9 = 9.
Shaun Dhuv . .	G.	445	4	4	Dhuv, 89 = 8.
Shaun Rhu . .	C.	570	3	4	Rhu, 214 = 7.
Sheela . . .	M.	341	8	6	
Sheerness . .	G.	630	9	6	
Sheet Anchor . .	H.	991	1	A.	Anchor, 281 = 2.
Sheffield Blade .	C.	480	3	2	Blade, 46 = 1.
Shekels . . .	F.	420	6	2	
Shellmartin . .	C.	1031	5	3	
Shenfield . . .	C.	484	7	3	
Shepherd King .	C.	684	9	4	King, 90 = 9.
Shepherd Lord .	C.	830	2	4	Lord, 236 = 2.
Sheridan . . .	G.	564	6	4	
Sheriff Hutton .	G.	1045	1	A.	Hutton, 455 = 5.
Sherwood . .	G.	520	7	A.	
Shevian . . .	F.	450	9	4	
Shifter . . .	M.	980	8	6	
Shillitoe . . .	C.	345	3	2	
Shillmoor . . .	G.	576	9	4	
Shilly Shally . .	G.	681	6	A.	
Shinju . . .	C.	359	8	2	
Ship-shape . .	G.	770	5	A.	
Shoda . . .		311	5	A.	
Short Circuit . .	C.	1582	7	3	Circuit, 680 = 5.

Name.	Sex.	Value.	Remainder.	Age (figures indicate years and "A" aged).	Remarks.
Short Gun . .	C.	772	7	3	Gun, 70=7.
Sicily Queen . .	M.	240	6	A.	Queen, 80=8.
Sidewing . . .	M.	140	5	6	
Sidus . . .	H.	124	7	5	
Silent Friend . .	C.	894	3	4	Friend, 344=2.
Silkhat . . .	G.	516	3	A.	Hat, 406=1.
Silly Girl . . .	M.	350	8	A.	Girl, 250=7.
Silver Bird . .	F.	576	9	4	Bird, 206=8.
Silver Bow . .	C.	378	9	4	Bow, 8=8.
Silver Bullet . .	F.	812	2	4	Bullet, 442=1.
Silverfield . .	H.	494	8	5	
Silverfort . . .	F.	1056	3	6	
Silverhampton . .	F.	866	2	4	
Silverrae . . .	C.	580	4	2	
Silver Sand II. .	M.	485	8	A.	
Silver Slipper . .	M.	662	5	5	Slipper, 292=4.
Silver Song . .	C.	502	7	3	Song, 132=6.
Silver Tariff . .	G.	1051	7	5	Tariff, 681=6.
Silver Valley . .	F.	491	5	4	Valley, 121=4.
Silviane . . .	F.	231	6	3	
Simon Glover . .	C.	480	3	4	Simon, 150=6.
Simonhatch . .	C.	159	6	2	
Simonness . .	F.	220	4	4	
Simonopus . .	F.	302	5	4	
Simonsbath . .	G.	618	6	6	
Simonsworth . .	F.	821	2	3	
Simontault (By) .	F.	982	1	3	
—— . . .	F.	982	1	3	
—— . . .	F.	982	1	2	
Simon the Cellarer .	G.	280	1	3	
Simony . . .	F.	160	7	3	
Simplon . . .	C.	260	8	3	
Single Shot . .	G.	862	7	4	Shot, 702=9.
Singlestick . .	C.	640	1	3	Stick, 480=3.
Sing On . . .	F.	182	2	3	
Sinopi . . .	G.	206	8	6	
Sinople . . .	C.	226	1	4	
Sirdar . . .	H.	465	6	A.	
Sir David . .	G.	275	5	A.	David, 15=6.
Sir Davy . . .	C.	355	4	3	Davy, 95=5.
Sir Francis Drake .	H.	945	9	A.	Drake, 234=9.
Sir Fretful . .	H.	1060	7	5	Fretful, 800=8.
Sir George . .	G.	468	9	4	George, 208=1.
Sir Hassard . .	H.	530	8	A.	Hassard, 270=9.

Name.	Sex.	Value.	Remainder.	Age (figures in-dicate years and "A" aged).	Remarks.
Sir Hugo (By) . . .	C.	307	1	2	Hugo, 47=2.
—— . . .	F.	307	1	2	Hugo, 47=2.
Sir Joshua . . .	C.	572	7	2	Joshua, 312=6.
Sir Pat . . .	C.	741	3	3	Pat, 481=4.
Sir Peter . . .	G.	950	5	A.	Peter, 690=6.
Sir Peter Lely . .	C.	1030	4	3	Lely, 80=8.
Sir Theo . . .	G.	681	6	4	Theo, 421=7.
Sir Vassar . . .	G.	601	7	A.	Vassar, 341=8.
Sir Visto (By) . .	F.	415	1	2	Visto, 155=2.
—— . . .	C.	415	1	2	Visto, 155=2.
—— . . .	F.	415	1	2	Visto, 155=2.
—— . . .	C.	415	1	2	Visto, 155=2.
Sir Walter . . .	G.	897	6	5	Walter, 637=7.
Sister Olive . . .	F.	832	4	3	Olive, 112=4.
Sister Sarah . .	F.	982	1	4	Sarah, 262=1.
Sister Superior . .	M.	1280	2	6	Superior, 560=2.
Sister to Addio . .	F.	1147	4	2	Addio, 21=3.
Skin Deep . . .	F.	224	8	2	
Skinful . . .	G.	240	6	4	
Skiograph . . .	C.	397	1	2	
Skip Jack . . .	G.	184	4	A.	Jack, 24=6.
Skirlnaked . . .	G.	394	7	A.	
Skylight . . .	F.	520	7	3	
Skyscraper . . .	F.	660	3	2	
Slander . . .	F.	345	3	3	
Slander II. . .	G.	345	3	A.	
Slave Driver . .	G.	664	7	4	Driver, 484=7.
Sleeping Beauty .	M.	678	3	5	Beauty, 428=5.
Sleepless . . .	F.	280	1	4	
Sleepy . . .	F.	190	1	3	
Slemish . . .	H.	440	8	A.	
Slieverue . . .	G.	386	8	5	
Slingsby . . .	G.	232	7	A.	
Slipburn . . .	G.	422	8	4	
Slipper . . .	F.	292	4	3	
Slogan . . .	F.	166	4	3	
Slogger . . .	G.	312	6	4	
Slowburn . . .	H.	348	6	6	
Sly Boots . . .		568	1	5	Boots, 468=9.
Sly Fox . . .	H.	292	4	A.	Fox, 192=3.
Small Carbine . .	G.	414	9	3	Carbine, 283=4.
Small Crop . .	F.	433	1	2	Crop, 302=5.
Smike . . .	G.	130	4	A.	
Smilare . . .	F.	211	4	2	

Name.	Sex.	Value.	Remainder.	Age (figures indicate years and "A" aged).	Remarks.
Smiling Morn . .	C.	492	6	4	Morn, 292 = 4.
Smiler . . .	G.	330	6	2	
Smokeless .	M.	226	1	5	
Smoking Concert .	G.	928	1	4	Concert, 732 = 3.
Snaffle . . .	G.	221	5	2	
Snake in the Grass .	M.	486	9	5	Snake, 140 = 5.
Snap . . .	G.	191	2	A.	
Snape . . .	G.	200	2	A.	
Snapshot . .	M.	893	2	5	
Snarley Yow .	H.	367	7	A.	Snarley, 351 = 9.
Sneeze .	G.	127	1	5	
Sniovadoir .	C.	418	4	3	
Snip . . .	F.	190	1	3	
Snipe . . .	M.	200	2	6	
Snowberry .	C.	338	5	3	
Snowden .	G.	170	8	6	
Snowdrift II. .	M.	800	8	A.	
Snowdrop .	F.	402	6	2	
Sobraon . .	G.	325	1	4	
Socks II. .	G.	142	7	5	
Socrates . .	C.	752	5	3	
Sohemus .	G.	181	1	4	
Soldanella .	F.	192	3	3	
Solera . .	F.	307	1	2	
Solicitor . .	C.	752	5	4	
Solidago . .	G.	123	6	5	
Soloman (By) .	M.	188	8	5	
Soloman II. .	G.	188	8	A.	
Somerlad . .	G.	350	8	3	
Something II. .	G.	575	8	A.	
Something Hot .	H.	982	1	6	Hot, 407 = 2.
Somnambulist II. .	M.	649	1	5	
Sonatura . .	M.	724	4	5	
Song of the Wood .	M.	237	3	A.	Song, 132 = 6.
Sonnetta . .	F.	136	1	3	Son, 110 = 2.
Son of the Morning .	G.	567	9	6	
Sonoma . .	C.	157	4	3	
Sooner . .	G.	316	1	5	
Sorceress . .	M.	596	2	5	
Sorceress . .		596	2	3	
Sorciere . .	F.	337	4	4	
Sospello . .	H.	248	5	5	
Soubrette III. .	M.	678	3	A.	
Soudanese II. .	M.	138	3	5	

Name.	Sex.	Value.	Remainder.	Age (figures indicate years and "A" aged).	Remarks.
Sour Cherry	G.	489	3	6	Cherry, 223 = 7.
Southern Cross	G.	596	2	A.	Cross, 282 = 3.
Southerness	H.	384	6	A.	
Southern Sea	M.	384	6	6	Sea, 70 = 7.
Sovereign	F.	402	6	4	
Spade Guinea	C.	234	9	3	Guinea, 80 = 8.
Spado	G.	151	7	A.	
Spanish Main	H.	591	6	5	Main, 100 = 1.
Spark II.	G.	361	1	A.	
Sparrow	C.	347	5	2	
Sparsholt	G.	1077	6	6	
Spatchcock	G.	186	6	4	
Spearwort	F.	956	2	2	
Spec.	F.	170	8	4	
Speciality	G.	891	9	A.	
Speckles	F.	260	8	3	
Spectrum	M.	810	9	6	
Specularia	F.	428	5	4	
Speculation II.	G.	567	9	A.	
Speculator	G.	817	7	3	
Speedy	G.	164	2	5	
Spendthrift	F.	1289	2	A.	
Sperate	F.	760	4	4	
Spiddal	G.	174	3	6	
Spider	G.	354	3	A.	
Spider II.	G.	354	3	A.	
Spider III.	G.	354	3	A.	
Spillane	G.	230	5	A.	
Spin.	C.	190	1	2	
Spinnaker II.	G.	411	6	6	
Spinning Boy	G.	274	4	A.	Boy, 14 = 5.
Spinning Top	G.	742	4	6	Top, 482 = 5.
Spion	G.	201	3	6	
Spirited	F.	754	7	4	
Spitfire	M.	830	2	A.	
Spitfire II.	M.	830	2	A.	
Splash Point	G.	1003	4	3	Point, 532 = 1.
Splice	F.	230	5	4	
Splinter	H.	820	1	5	
Splutter	H.	770	5	5	
Spoonley	G.	236	2	A.	
Sportsman II.	G.	896	5	A.	
Sportsman III.	G.	896	5	A.	
Spotted Dog	G.	592	7	A.	Dog, 26 = 8.

Name.	Sex.	Value.	Remainder.	Age (figures indicate years and "A" aged).	Remarks.
Spratty . . .	G.	751	4	3	
Spread Eagle . .	M.	415	1	A.	Eagle, 61 = 7.
Sprig III. . .	G.	360	9	A.	
Sprig of Shillelagh .	H.	812	2	5	Spring, 360 = 9.
Spring Boy . .	C.	424	1	3	Boy, 14 = 5.
Springbuck . .	G.	432	9	A.	Buck, 22 = 4.
Spring Duke . .	G.	450	9	4	Duke, 40 = 4.
Springfield . .	G.	534	3	A.	
Spring Flower . .	H.	728	8	6	Flower, 318 = 3.
Spring Hare . .	H.	625	4	5	Hare, 215 = 8.
Springlock . .	H.	462	3	6	
Spring Mart . .	C.	1051	7	3	Mart, 641 = 2.
Spring Meeting . .	G.	930	3	4	Meeting, 520 = 7.
Springvale . .	F.	530	8	4	
Spring Weather . .	H.	630	2	6	Weather, 220 = 4.
Spring Witch . .	M.	419	5	5	Witch, 9 = 9.
Square . . .	G.	290	2	5	
Squeeze II. . .	M.	290	2	A.	
Squid . . .	F.	84	3	2	
Squint II. . .	M.	530	8	5	
Squire . . .	H.	290	2	6	
Squire Jack . .	G.	314	8	A.	Jack, 24 = 6.
Stagsden . . .	F.	595	1	3	
Stalker . . .	G.	682	7	A.	
Standard . . .		719	8	A.	
Stansted . . .	C.	985	4	3	
Starch . . .	M.	664	7	A.	
Stargazer . . .	G.	898	7	2	
Starlight III. (By) .	F.	1091	2	2	
Starling . . .	M.	761	5	A.	
Star of Hanover .	H.	1084	4	5	Star, 661 = 4.
Star of Hope . .	H.	833	5	6	Star, 661 = 4.
Star of Thirsk . .	F.	1427	5	4	Star, 661 = 4.
Stars and Stripes .	C.	1486	1	3	Stars, 721 = 1.
Starslee . . .	M.	761	5	A.	
Start Bay . . .	G.	1073	2	6	Bay, 12 = 3.
Startling . . .	F.	1161	9	3	
Statesman II. . .	G.	1020	3	5	
Stealaway . . .	G.	517	4	5	
Steady Glass . .	G.	595	1	A.	Glass, 111 = 3.
Stedalt . . .	G.	905	5	A.	
Stella III. . .	M.	501	6	6	
Step Child . . .	F.	587	2	2	Child, 37 = 1.
Step Forward . .	G.	1042	7	5	Forward, 492 = 6.

Name.	Sex.	Value.	Remainder.	Age (figures indicate years and "A" aged).	Remarks.
Stephano . . .	G.	607	4	4	
Stephen . . .	G.	600	6	A.	
Stephens . . .	G.	660	3	A.	
Stepping Stone .	M.	1058	5	A.	Stone, 516 = 3.
Sterling Balm .	F.	803	2	3	Balm, 44 = 8.
Sterling Bank (By) .	F.	833	5	2	Bank, 73 = 1.
Sterling Blue .	F.	798	6	2	
Sternberg . .	G.	932	5	A.	
Stevedore . .	G.	760	4	A.	
Steventon . .	G.	1050	6	6	
Stewardess . .	F.	750	3	2	
Stick in the Mud .	M.	589	4	6	Stick, 480 = 3.
Still	F.	490	4	4	
Sting II. . . .	G.	530	8	A.	
Stirtloe . . .	G.	1096	7	A.	
Stockpot . .	F.	964	1	2	
Stodart's Blend .	H.	1227	3	6	Blend, 96 = 6.
Stolen Bride .	M.	752	5	5	Bride, 206 = 8.
Stormcock . .	G.	744	6	6	
Stormfiend . .	G.	846	9	A.	
Storm Signal .	G.	862	7	A.	Signal, 160 = 7.
Stornoway . .	G.	734	5	6	
Straight On .	G.	1122	6	5	Straight, 1070 = 8.
Strangford . .	G.	1016	8	6	
Stratheden . .	G.	1130	5	6	
Strategus (By) .	G.	1151	8	2	
Strategy . .	F.	1084	4	3	
Strathspey .	M.	1216	1	6	
Strawberry Leaf .	F.	1004	5	4	Leaf, 120 = 3.
Stray Star .	G.	1331	8	6	Star, 661 = 4.
Street Lamp .	H.	1221	6	5	Lamp, 151 = 7.
Strelma . .	H.	741	3	6	
Streptocarpus .	M.	1517	5	5	
Stretton . .	C.	1120	4	3	
Strideaway .	G.	681	6	5	
Stroller . .	G.	896	5	A.	
Strontian . .	G.	1062	9	4	
Studding Sail .	G.	634	4	A.	Sail, 100 = 1.
Sub Rosa . .	M.	276	6	5	
Success . .	M.	210	3	5	
Successful . .	G.	320	5	A.	
Succotash . .	G.	787	4	4	
Sudd . . .	C.	64	1	4	
Sudden Rise .	F.	321	6	2	

Name.	Sex.	Value.	Remainder.	Age (figures indicate years and "A" aged).	Remarks.
Suezala . . .	M.	109	1	5	
Suezath . . .	F.	483	6	3	
Summerhill . .	G.	335	2	6	
Summer Shower .	F.	808	7	4	
Sunbeam III. .	M.	162	9	6	
Sun Bonnet . .	F.	574	7	5	Bonnet, 464 = 5.
Sunburnt . . .	F.	762	6	2	
Sundew . . .	F.	130	4	2	
Sun Dog . . .	G.	136	1	4	Dog, 26 = 8.
Sundorne . . .	H.	370	1	5	
Sundridge . .	C.	317	2	4	
Sunflower . .	F.	428	5	2	
Sunny Shower .	G.	628	7	A.	Shower, 508 = 4.
Sunny South .	F.	591	6	2	South, 471 = 3.
Sun Rose . . .	F.	323	8	2	Rose, 213 = 6.
Sunset . . .	F.	580	4	4	
Superbus . . .	G.	402	6	3	
Surcouf . . .	C.	366	6	3	
Surefoot (By) .	F.	986	5	3	
Surefoot II. .	G.	986	5	A.	
Surehaven . .	M.	642	3	A.	
Surf Duck . .	M.	364	4	A.	Duck, 24 = 6.
Surprenant . .	C.	1051	7	3	
Surprise . .		547	7	6	
Surprise Hill .	G.	582	6	6	Hill, 35 = 8.
Survivor . . .	H.	620	8	A.	
Susanna . . .	F.	172	1	2	
Suttamore . .	C.	707	5	4	
Suzanne . . .	F.	118	1	3	
Suzanne II. .	M.	118	1	5	
Svelte . . .	F.	580	4	3	
Swaledale . .	G.	180	9	A.	
Swallow . . .	G.	133	7	A.	
Sweet Briar .	G.	909	9	4	Briar, 403 = 7.
Sweet Charlotte .	M.	1140	6	A.	Charlotte, 634 = 4.
Sweet Clorane .	F.	822	3	2	Clorane, 316 = 1.
Sweet Dixie .	F.	630	9	4	Dixie, 124 = 7.
Sweetheart (By)	F.	1112	5	3	
——— . .	F.	1112	5	3	
Sweetheart .	M.	1112	5	A.	
Sweetheart III. .	M.	1112	5	6	
Sweet Hilda .	F.	546	6	4	Hilda, 40 = 4.
Sweet Incense .	M.	737	8	6	Incense, 231 = 6.
Sweet Kirconnel	F.	838	1	3	Kirconnel, 332 = 8.

Name.	Sex.	Value.	Remainder.	Age (figures indicate years and "A" aged).	Remarks.
Sweet Kiss	C.	586	1	2	Kiss, 80=8.
Sweet Marion	M.	807	6	A.	Marion, 301=4.
Sweetmore	G.	751	4	4	
Sweet Mountain	C.	1052	8	4	Mountain, 546=6.
Sweet Nell	M.	596	2	5	Nell, 90=9.
Sweet Patti	M.	997	7	5	Patti, 491=5.
Sweet Oliver	C.	818	8	4	Oliver, 312=6.
Sweet Repose	F.	809	8	3	Repose, 303=6.
Sweet Rose	F.	719	8	2	Rose, 213=6.
Sweets	F.	566	8	3	
Sweet Singer	M.	836	8	6	Singer, 330=6.
Sweet Sounds	H.	686	2	5	Sounds, 180=9.
Sweet Ulva	F.	618	6	4	Ulva, 112=4.
Sweet Viola	G.	633	3	A.	Viola, 127=1.
Sweetwort	G.	1112	5	A.	
Swell II.	G.	136	1	A.	
Swift	H.	576	9	5	
Swift Cure	F.	812	2	2	Cure, 236=2.
Swillington (By)	F.	646	7	3	
Swiss Girl	F.	406	1	3	Girl, 250=7.
Switchback	F.	122	5	4	
Switch Cap	F.	200	2	2	Cap, 101=2.
Swona	F.	147	3	2	
Sybarite	H.	663	6	A.	
Syerla	H.	301	4	A.	
Sylvan Park	F.	448	7	3	Park, 301=4.
Sylvine	M.	230	5	5	
Syme	C.	110	2	2	
Symington	F.	620	8	2	
Syndicate II.	G.	544	4	A.	
Syneros	C.	382	4	4	
Table Tennis	G.	953	8	A.	Tennis, 520=7.
Taboo	G.	409	4	2	
Taffrail	G.	721	1	A.	
Tait's Clock	F.	942	6	4	Clock, 72=9.
Tamar	F.	641	2	2	
Tambour	G.	648	9	A.	
Tame Fox	G.	642	3	A.	Fox, 192=3.
Tamworth	G.	1052	8	A.	
Tankerness	G.	791	8	5	
Tantalus	C.	942	6	4	
Tanzmeister (By)	F.	1168	7	2	
————	F.	1168	7	2	

13

Name.	Sex.	Value.	Remainder.	Age (figures indicate years and "A" aged).	Remarks.
Tanzmeister (By) .	F.	1168	7	3	
Tapestry . . .	M.	1151	8	6	
Taplow . . .	C.	517	4	4	
Tarara . . .	F.	802	1	2	
Tarasp . . .	F.	742	4	4	
Tarmon Valley. .	G.	812	2	6	Valley, 121 = 4.
Tarnbrook . .	F.	879	6	2	
Tarolinta . .	M.	1088	8	6	
Tarporley (By) .	F.	927	9	2	
——— . . .	F.	927	9	2	
——— . . .	C.	927	9	3	
——— . . .	F.	927	9	2	
——— . . .	F.	927	9	2	
——— . . .	F.	927	9	2	
——— . . .	C.	927	9	3	
Taporley Knight .		1377	9	5	Knight, 450 = 9.
Tarquin . . .	G.	671	5	5	
Tarsney . . .	G.	721	1	4	
Tascara . . .	G.	683	8	2	
Taskmaster . .	C.	1182	3	3	Master, 701 = 8.
Taste . . .	F.	870	6	3	
Tatcho II. . .	M.	410	5	A.	
Tatiana . . .	F.	863	8	4	
Tatius . . .	G.	761	5	3	
Tavennes . . .	C.	661	4	2	
Tavora . . .	G.	688	4	A.	
Tea Cosy . . .	F.	453	3	4	Cosy, 45 = 9.
Teddesley . .	C.	514	1	4	
Teetotum . . .	F.	1256	5	3	
Telegram . . .	G.	711	9	A.	
Tell Tale . . .	F.	880	7	4	
Teltown . . .	G.	896	5	6	
Templemore . .	C.	806	5	4	
Tempo . . .	G.	536	5	5	
Temptation II. .	G.	1201	4	6	
Tenebrosa . .	G.	739	1	5	
Teredo . . .	H.	638	9	A.	
Testbourne . .	G.	1128	3	2	
Testwood . . .	G.	880	7	2	
Teuton . . .	C.	866	2	4	
Teviot II. . .	H.	900	9	A.	
Thanks II. . .	M.	536	5	A.	
Thanksgiving . .	F.	632	2	2	
Thelpusa . . .	M.	586	1	6	

Name.	Sex.	Value.	Remainder.	Age (figures indicate years and "A" aged).	Remarks.
Theodosia	F.	497	2	4	
Theodocion	C.	551	2	4	
Theopathy	F.	912	3	2	
Theophilus (By)	G.	591	6	3	
Theorbo	C.	629	8	3	
Theorist	C.	1081	1	3	
Theorist II.	G.	1081	1	A.	
Thessalia	F.	517	4	2	
Thick Fog	C.	527	5	3	
Thirlstane	H.	1155	3	A.	
Thisbe	F.	477	9	4	
Thistledown	M.	555	6	A.	
Thoas	G.	472	4	3	
Thomondgate	G.	931	4	5	
Thornton	G.	1107	9	5	
Thoughtless	M.	907	7	A.	
Thranum	G.	706	4	6	
Three Star	G.	1276	7	A.	Star, 661 = 4.
Thremhall	G.	692	8	4	
Thrift	F.	1085	5	2	
Throwaway	C.	628	7	3	
Thuja	M.	409	4	5	
Thumbawn	G.	499	4	A.	
Thunia	F.	466	7	3	
Thunderbolt	C.	1097	8	2	
Thunderclap	G.	790	7	A.	
Thurlby	G.	647	8	5	
Thurifer	G.	885	3	4	
Thursday II.	C.	626	5	3	
Thyases	H.	546	6	5	
Tiboro	F.	614	2	5	
Tight Fit	G.	1280	2	A.	
Tiltonette	F.	905	5	4	
Timbre	C.	642	3	3	
Time	C.	440	8	4	
Time Fuse	M.	543	3	5	Fuse, 103 = 4.
Timely Warning	M.	807	6	5	Warning, 327 = 3.
Timeserver	G.	980	8	2	
Time Spinner	F.	830	2	3	Spinner, 390 = 3.
Timon	H.	490	4	A.	
Timpout	M.	926	8	6	
Tinfein	F.	590	5	2	
Tin Soldier	G.	749	2	4	Soldier, 299 = 2.
Tintagel Castle	C.	995	5	4	Castle, 111 = 3.

Name.	Sex.	Value.	Remainder.	Age (figures indicate years and "A" aged).	Remarks.
Tintara . . .	G.	1052	8	A.	
Tintara II. . .	M.	1052	8	A.	
Tiny . . .	F.	460	1	4	
Tip Cat . . .	M.	901	1	A.	Cat, 421 = 7.
Tippler . . .	C.	640	1	2	
Tipperary Boy . .	H.	827	8	A.	Boy, 14 = 5.
Titchfield . . .	G.	527	5	A.	
Titus II. . . .	G.	860	5	6	
Tit-willow . .	G.	842	5	A.	
Toastmaker . .	M.	1127	2	6	
Tobasco II. . .	G.	495	9	4	
Toddy . . .	F.	416	2	2	
Tod Sloan . .	H.	552	3	5	
Tokenhouse . .	G.	547	7	2	
Tollburn . . .	C.	694	1	3	
Tom . . .	C.	442	1	4	
Tomboy II. . .	F.	456	6	4	Boy, 14 = 5.
Tommy . . .	G.	452	2	A.	
Tommy Atkins II. .	G.	993	3	A.	Tommy, 452 = 2.
Tommy Atkins .	G.	993	3	A.	Tommy, 452 = 2.
Tom Pinch . .	G.	575	8	A.	Tom, 442 = 1.
Tom Tit . . .	G.	1242	9	6	Tom, 442 = 1.
Tomtit II. . .	G.	872	8	A.	
Tomtit III. . .	G.	872	8	A.	
Tom Tucker . .	C.	1062	9	3	Tom, 442 = 1.
Tonsure . . .	G.	958	4	4	
Tony . . .	G.	462	3	3	
Too Early . .	G.	647	8	2	Early, 241 = 7.
Too Quick . .	F.	446	5	4	Quick, 40 = 4.
Topo . . .	F.	492	6	3	
Topsail II. . .	M.	582	6	6	
Toronto . . .	G.	1064	2	4	
Torpedo III. . .	G.	702	9	A.	
Torrent . . .	G.	1062	9	3	
Tortion . . .	G.	956	2	5	
Tory . . .	G.	616	4	6	
—— . . .	G.	616	4	6	
Total Terror . .	F.	1646	8	3	Terror, 810 = 9.
Tourin . . .	G.	656	8	4	
Tours . . .	G.	666	9	A.	
Tower Hill . .	F.	643	4	4	Hill, 35 = 8.
Tramp . . .	G.	721	1	6	
Tramp . . .	H.	721	1	A.	
Tranquillity . .	F.	1111	4	4	

Name.	Sex.	Value.	Remainder.	Age (figures indicate years and "A." aged).	Remarks.
Transparency .	F.	1122	6	4	
Trappist . .	G.	1063	1	A.	
Trastevere . .	C.	1361	2	3	
Traveller II. . .	H.	911	2	5	
Travieso . . .	H.	757	1	A.	
Traynor . . .	G.	860	5	A.	
Tredennis . .	C.	734	5	4	
Trefoil II. . .	F.	722	2	4	
Trelydan . . .	M.	705	3	5	
Trenchant . .	C.	1114	7	2	
Trenton (By) . .	C.	1110	3	2	
—— . . .	C.	1110	3	2	
—— . . .	C.	1110	3	2	
Trenton Falls . .	F.	1281	3	2	
Trentonita . .	M.	1517	5	6	
Trevor . . .	H.	890	8	A.	
Tricky Devil . .	G.	754	7	2	Devil, 124 = 7.
Triolet . . .	F.	1056	3	2	
Trivia . . .	F.	691	7	3	
Trivial . . .	M.	720	9	5	
Troglodyte . .	F.	1072	1	3	
Troubadour (By) .	F.	819	9	3	
Trouvere . . .	G.	896	5	5	
True Blue . .	M.	644	5	6	
Truefoot . . .	G.	1086	6	4	
Truehaven . .	C.	742	4	3	
Trueno . . .	G.	672	6	A.	
Trustee . . .	G.	1070	8	5	
Trusty . . .	G.	1070	8	4	
Try Again . .	M.	691	7	A.	Try, 610 = 7.
Try On . . .	M.	662	5	5	Try, 610 = 7.
Tryst . . .	M.	1060	7	A.	
Tube Rose . .	M.	631	1	5	Rose, 213 = 6.
Tucka Tucka . .	C.	842	5	3	Tucka, 421 = 7.
Tudor King . .	C.	710	8	3	Tudor, 620 = 8; King, 90 = 9.
Tully Lass . .	F.	531	9	2	Lass, 91 = 1.
Tumulus . . .	C.	530	8	3	
Tune . . .	F.	466	7	2	
Tunnel Section .	G.	920	2	A.	Section, 440 = 8.
Turban II. . .		652	4	6	
Turf Lodge . .	C.	715	4	4	
Turk II. . . .	G.	620	8	A.	
Turtle Dove . .	F.	1114	7	4	Dove, 84 = 3.

Name.	Sex.	Value.	Remainder.	Age (figures indicate years and "A" aged).	Remarks.
Turveydrop . .	C.	976	4	2	
Tuscan . . .	G.	530	8	2	
Tuscapeda . .	M.	570	9	6	
Tuskar . . .	H.	681	6	5	
Tutti Frutti . .	F.	1500	6	2	Tutti, 810=9.
Twice Shy . .	M.	385	7	6	Shy, 310=4.
Twilight . . .	M.	455	5	6	
Twitchbell . .	F.	60	6	2	
Twitter . . .	F.	615	3	2	
Two	F.	406	1	4	
Tyna . . .	M.	451	1	A.	
Tyneholme . .	G.	501	6	6	
Tyninghame . .	C.	575	8	4	
Tynwald Hill .	C.	526	4	2	Hill, 35=8.
Typical Balsam .	F.	663	6	3	Balsam, 133=7.
Tyrant (By) . .	C.	1050	6	2	
—— . . .	F.	1050	6	2	
—— . . .	C.	1050	6	3	
Tyro . . .	G.	606	3	A.	
Tyrrhenian . .	G.	725	5	5	
Ugolino . . .	H.	118	1	6	
Ulster Boy . .	C.	705	3	4	
Ulterior . . .	H.	851	5	A.	
Una	F.	67	4	2	
Unanina . .	F.	162	9	4	
Uncanonical .	F.	228	3	3	
Uncle Charlie .	G.	345	3	A.	Charlie, 244=1.
Uncle George .	G.	309	3	A.	George, 208=1.
Uncle Henry .	G.	376	7	5	Henry, 275=5.
Uncle Jim . .	H.	144	9	5	Jim, 43=7.
Uncle Mac . .	G.	162	9	A.	Mac, 61=7.
Uncle Sam .	G.	202	4	A.	Sam, 101=2.
Undecided (By) .	G.	143	8	2	
—— . .	G.	143	8	2	
Underbred . .	H.	471	3	6	
United States .	C.	1404	9	3	States, 930=3.
Unknown . .	G.	157	4	A.	
Unsightly . .	F.	551	2	4	
Untameable . .	C.	534	3	2	
Upcot . . .	M.	503	8	6	
Upper Cut . .	G.	623	2	A.	
Upstart (By) . .	C.	1142	8	2	
Urfa	F.	282	3	2	

Name.	Sex.	Value.	Remainder.	Age (figures indicate years and "A" aged).	Remarks.
Ursula	F.	292	4	4	
Urugayo	H.	253	1	6	
Useful Boy	H.	200	2	A.	Boy, 14 = 5.
Vagabond	C.	160	7	3	
Vagrant II.	C.	751	4	2	
Vain Pride	F.	424	1	2	Pride, 284 = 5.
Val de Saire	C.	395	8	3	Val, 111 = 3.
Valdis	G.	175	4	6	
Valenza	F.	179	8	4	
Valhalla	M.	148	4	A.	
Valiant	C.	571	4	3	
Validity	G.	525	3	3	
Valise	F.	181	1	2	
Valla	G.	112	4	3	
Van	C.	131	5	4	
Vandam	G.	176	5	4	
Vandevelde	C.	279	9	3	
Van Houten	G.	592	7	A.	Houten, 461 = 2.
Vanishing Lady	F.	546	6	4	Lady, 45 = 9.
Varna	F.	332	8	3	
Vassar Miss	M.	441	9	6	Miss, 100 = 1.
Vasto	C.	156	3	4	
Vatel	H.	521	8	A.	
Vates	C.	550	1	4	
Vaubau II.	C.	104	5	3	
Veiled Queen	M.	204	6	5	Queen, 80 = 8.
Veles	C.	180	9	4	
Velum	C.	160	7	3	
Velutina	F.	571	4	3	
Vence	C.	200	2	4	
Vendale	C.	184	4	3	
Vendetta		554	6	3	
Veneer	C.	350	8	3	
Venerable Bede	G.	389	2	A.	Bede, 16 = 7.
Venetia	F.	451	1	2	
Venetian Belle	F.	542	2	4	Belle, 42 = 6.
Venetian Monk	G.	612	9	6	Monk, 112 = 4.
Ventilator II.	G.	1171	1	4	
Venture II.	M.	343	1	6	
Venturesome	C.	443	2	3	
Venus	F.	200	2	3	
Veracity (By)	F.	761	5	3	
——	G.	761	5	3	

Name.	Sex.	Value.	Remainder.	Age (figures indicate years and "A" aged).	Remarks.
Veracity (By) . .	G.	661	5	2	
Verdi . . .	C.	294	6	4	
Verglass . . .	C.	391	4	2	
Veridian . . .	F.	354	3	4	
Veritable . . .	F.	723	3	3	
Veritas . . .	H.	751	4	5	
Vert de Gris . .	C.	924	6	3	Vert, 680 = 5.
Vertigo . . .	G.	706	4	3	
Verus . . .	C.	400	4	3	
Vesta II. . .	M.	551	2	A.	
Veturia . . .	M.	707	5	4	
Vibrant . . .	C.	733	4	2	
Vibrate (By) . .	F.	692	8	3	
Vicator . . .		700	7	A.	
Vickers . . .	C.	360	9	3	
Victor Don . .	H.	756	9	A.	Victor, 700 = 7.
Victor Hugo . .	H.	747	9	5	Hugo, 47 = 2.
Victoria Cross .	M.	999	9	6	Cross, 282 = 3.
Victoria Hill .	F.	752	5	2	Hill, 35 = 8.
Victor Lancefield .	G.	965	2	3	Victor, 700 = 7.
Vidame . .	G.	134	8	4	
Vide Bouteilles .	C.	592	7	3	Vide, 84 = 3.
Vieille Moule .	M.	196	7	5	Vieille, 120 = 3
Vier Marchi .	C.	544	4	3	Vier, 290 = 2.
Vieux Marchem	C.	687	3	4	Vieux, 96 = 6.
View Holla .	H.	134	8	5	View, 96 = 6 ; Holla, 38 = 2.
Village Beau .	C.	131	5	3	Beau, 8 = 8.
Villager II. .	G.	323	8	A.	
Villikins . .	C.	240	6	4	
Vinaigrette .		740	2	6	
Vincent . .	G.	640	1	A.	
Vine Leaf .	G.	260	8	5	Leaf, 120 = 3.
Vineyard . .	G.	355	4	5	
Violent . .	F.	586	1	2	
Violet . .	M.	536	5	A.	
Violet Ethel .	F.	982	2	4	Violet, 536 = 5.
Violetta . .	M.	537	6	5	
Viper . .	G.	360	9	4	
Vision . .	F.	137	2	4	
Visionary . .	G.	347	5	3	
Visor . .	C.	342	9	2	
Vittel . .	F.	510	6	4	
Viva II. . .	M.	161	8	6	

Name.	Sex.	Value.	Remainder.	Age (figures indicate years and "A" aged).	Remarks.
Vive le Roi . . .	G.	422	8	3	Roy, 212 = 5.
Vixen II. . . .	M.	210	3	A.	
Vogelkop . . .	G.	238	4	3	
Voggin . . .	G.	156	3	5	
Volatile . . .	F.	557	8	4	
Volenite . . .	G.	576	9	6	
Volodyowski . .	C.	222	6	4	
Volonel . . .	G.	208	1	4	
Volpone . . .	C.	248	5	4	
Volsinian . . .	H.	282	3	A.	
Vonitza . . .	F.	544	4	3	
Vulpia . . .	H.	206	8	5	
Wabun . . .	C.	59	5	4	
Wagner . . .	H.	277	7	5	
Wagon Hill . .	G.	112	4	4	Hill, 35 = 8; Wagon, 77 = 5.
Wagram . . .	M.	267	6	5	
Wag-Wag . . .	F.	54	9	3	
Wahrheit . . .	F.	622	1	2	
Wait a While . .	M.	463	4	A.	Wait, 416 = 2.
Wakefield (By) .	F.	160	7	2	
Walk Over . .	C.	314	8	4	Walk, 27 = 9.
Walnut . . .	H.	487	1	A.	
Walter Scott . .	G.	1119	3	6	Walter, 637 = 7.
Walwyn . . .	G.	93	3	A.	
Wandering Minstrel	G.	1111	4	3	Minstrel, 780 = 6.
Warbird . . .	M.	413	8	A.	
War Cloud . .	G.	267	6	6	Cloud, 60 = 6.
War Game . .	G.	277	7	5	Game, 70 = 7.
Wargrave . . .	C.	517	4	4	
Warhoop . . .	G.	298	1	A.	
War Kit . . .	C.	627	6	4	Kit, 420 = 6.
Warlaby (By) . .	C.	249	6	2	
—— . . .	G.	249	6	4	
Warlock . . .	G.	259	7	A.	
Warm Baths . .	G.	715	4	4	Baths, 468 = 9.
Warminster . .	C.	957	3	3	Minster, 750 = 3.
Warning . . .	C.	327	3	4	
Warning Note . .	H.	783	9	6	Note, 456 = 6.
War Paint . .	F.	747	9	3	Paint, 540 = 9.
Warrenpoint . .	C.	789	6	4	
Warsop . . .	F.	349	7	3	
War Wolf . . .	C.	323	8	3	Wolf, 116 = 8.

Name.	Sex	Value.	Remainder.	Age (figures indicate years and "A" aged).	Remarks.
Watcher . . .	G.	210	3	6	
Water Baby . .	F.	622	1	3	Baby, 15 = 6.
Water Bird . .	M.	813	3	A.	Bird, 206 = 8.
Water Chute . .	C.	1313	8	3	Chute, 706 = 4.
Water Crow . .	C.	833	5	4	Crow, 226 = 1.
Water Shed . .	C.	921	3	4	Shed, 314 = 8.
Watertower . .	C.	1215	9	3	
Water Violet . .	F.	1143	9	4	Violet, 536 = 5.
Water Wheel . .	C.	653	5	3	Wheel, 46 = 1.
Water Wings . .	F.	743	5	3	Wings, 136 = 1.
Wave-less . .	C.	196	7	3	
Wavelets' Pride .	H.	880	7	5	Pride, 284 = 5.
Waxflower . .	F.	405	9	4	
Wax Toy . . .	H.	107	1	A.	Toy, 20 = 2.
Wearing o' the Green	F.	587	2	4	Wearing, 286 = 7.
Wedding March .	F.	334	1	4	March, 244 = 1.
Wedding Peal . .	F.	210	3	2	Peal, 120 = 3.
Wednesday II. .	M.	87	6	5	
Wee Busbie . .	G.	90	9	5	Busbie, 74 = 2.
Wee Janet . .	F.	480	3	2	Janet, 464 = 5.
Wee Nell . .	M.	106	7	6	Nell, 90 = 9.
Wees Kneppchen .	F.	498	3	2	Kneppchen, 422 = 8.
Well Done II.. .	G.	100	1	6	Done, 54 = 9.
Well Fort. . .	H.	728	8	6	Fort, 686 = 2.
Welsh Song . .	F.	478	1	3	Song, 132 = 6.
Wensleydale . .	C.	210	3	3	
Westcliffe . .	G.	606	3	A.	
West End . .	G.	541	1	2	End, 65 = 2.
West Malling . .	H.	617	5	5	Malling, 141 = 6.
Westminster (By) .	F.	1226	2	2	
Westmoreland . .	G.	807	6	5	
Weston . . .	G.	526	4	A.	
Westralian . .	H.	767	2	A.	
West Smithfield .	G.	1105	7	6	Smithfield, 629 = 8.
West Wind II. .	F.	536	5	4	Wind, 60 = 6.
Wexford Boy . .	C.	394	7	4	Boy, 14 = 5.
Whatkepou . .	F.	523	1	4	
What Next . .	C.	947	2	4	What, 407 = 2.
Wheel Lock . .	F.	98	8	4	Lock, 52 = 7.
Whiffingoffin . .	G.	308	2	4	
Whin Blossom . .	M.	190	1	6	Blossom, 134 = 8.
Whinflower . .	F.	374	5	3	
Whisht . . .	F.	706	4	3	
Whisper II. . .	M.	346	4	6	

Name.	Sex.	Value.	Remainder.	Age (figures indicate years and "A" aged).	Remarks.
Whisperer	G.	546	6	3	
Whistling Crow	C.	392	5	2	Crow, 226=1.
Whitebine	C.	468	9	4	
Whiteboy II.	G.	420	6	A.	Boy, 14=5.
White Eyes	G.	424	1	6	Eyes, 18=9.
White Frost	H.	1148	5	A.	Frost, 742=4.
Whitehall (By)	C.	443	2	2	
White Hart	G.	1012	4	A.	Hart, 606=3.
Whitehaven	G.	542	2	A.	
Whitehead II.	G.	425	2	A.	
White Nun II.	M.	506	2	6	Nun, 100=1.
White Peter	G.	1096	7	3	Peter, 690=6.
Whitesocks II.	G.	548	8	A.	
Whitesocks III.	F.	548	8	4	
Whitewash	G.	713	2	A.	
Whitby Bay	G.	430	7	3	Bay, 12=3.
Whittier (By)	F.	616	4	2	
Whittinghame	G.	521	8	3	
Wiedersehen	G.	302	5	4	
Wiki Wiki	F.	72	9	4	
Wild Apple	F.	83	2	3	Apple, 43=7.
Wild Bird	M.	246	3	6	Bird, 206=8.
Wild Boer	C.	248	5	2	Boer, 208=1.
Wild Fancy	C.	241	7	2	Fancy, 201=3.
Wild Fawn	M.	172	1	5	Fawn, 132=6.
Wildfire II.	G.	330	6	A.	
Wildfire III.	G.	330	6	A.	
Wild Flora	F.	357	6	2	Flora, 317=2.
Wild Flower	G.	358	7	A.	Flower, 318=3.
Wildfowl	F.	156	3	2	Fowl, 116=8.
Wildham	G.	85	4	4	
Wild Hopes	F.	191	2	4	Hopes, 151=7.
Wild Lad	C.	75	3	2	Lad, 35=8.
Wild Monk (By)	C.	152	8	2	Monk, 112=4.
Wild Night Again	F.	571	4	3	
Wild Novice	F.	232	7	4	Novice, 192=3.
Wild Nun	M.	140	5	6	Nun, 100=1.
Wild Seamew	F.	166	4	3	Seamew, 126=9.
Wild Sport	F.	786	3	2	Sport, 746=8.
Wild Weather	M.	260	8	6	Weather, 220=4.
Wilhelmina	F.	172	1	4	
William Bailey	G.	138	3	6	William, 86=5.
William the Third	C.	709	7	4	William, 86=5.
Willie	G.	46	1	A.	

Name.	Sex.	Value.	Remainder.	Age (figures indicate years and " A " aged).	Remarks.
Willox Hill . .	G.	153	9	A.	Hill, 35 = 8.
Wilton Castle . .	C.	597	3	4	Castle, 111 = 3.
Windowsill . .	G.	156	3	6	
Windsail . . .	H.	160	7	5	
Winkfield (By) .	G.	200	2	2	
—— . . .	C.	200	2	2	
—— . . .	C.	200	2	2	
Winkfield (By) .	G.	200	2	2	
Winkfield Junior .	G.	469	1	5	Winkfield, 200 = 2.
Winkfields Charm .	C.	504	9	3	Charm, 244 = 1.
Winkfields Dower .	G.	472	4	A.	Dower, 212 = 5.
Winkfields Pearl .	M.	570	3	6	Pearl, 310 = 4.
Winking Willie .	H.	192	3	4	Willie, 46 = 1.
Winnie . . .	F.	66	3	4	
Winnipeg . . .	M.	166	4	5	
Winsome Lad . .	G.	191	2	A.	Lad, 35 = 8.
Wiscorsin II. . .	C.	248	5	3	
Wise Bess . .	F.	85	4	4	Bess, 72 = 9.
Wise Lad . .	G.	48	3	5	Lad, 35 = 8.
Wise Morgan . .	H.	325	1	5	Morgan, 312 = 6.
Wiseman (By) . .	F.	104	5	2	
—— . . .	C.	104	5	2	
—— . . .	F.	104	5	2	
—— . . .	F.	104	5	2	
Wise Prince . .	C.	403	7	4	Prince, 390 = 3.
Wise Rosie . .	F.	236	2	3	Rosie, 223 = 7.
Wishaw . . .	G.	308	2	6	
Wishbone . .	F.	364	4	4	
Wisp . . .	C.	146	2	3	
Witch . . .	M.	9	9	A.	
Witch of the Hills .	M.	199	1	A.	Witch, 9 = 9.
Witticism . . .	G.	513	9	A.	
Witton Fell . .	C.	606	3	3	Witton, 486 = 9.
Witty Maid . .	F.	470	2	3	Maid, 54 = 9.
Wolf . . .	H.	116	8	5	
Wolfgang . .	G.	207	9	3	
Wolfsbane . .	F.	238	4	3	
Wolf's Crag (By) .	G.	364	4	2	Crag, 241 = 7.
—— . . .	F.	364	4	2	Crag, 241 = 7.
—— . . .	C.	364	4	2	Crag, 241 = 7.
Wolfshall . . .	C.	213	6	2	
Woodfield . .		134	8	A.	
Wood Pigeon II. .	G.	143	8	A.	Pigeon, 133 = 7.
Woollashill . .	G.	367	7	2	

Name.	Sex.	Value.	Remainder.	Age (figures indicate years and "A" aged).	Remarks.
Wooton Wood . .	G.	466	7	4	Wood, 10 = 1.
Worcester (By) .	F.	866	2	2	
Worcester (By) .	G.	866	2	4	
Working Man . .	C.	387	9	4	Man, 91 = 1.
Workmate . .	M.	676	1	6	
Worldly Wise (By) .	F.	263	2	2	Wise, 13 = 4.
Wrangler . .	G.	401	5	4	
Wyvern . . .	H.	336	3	A.	
Xenie . . .	M.	77	5	A.	
Xury (By) . .	C.	223	7	2	
Xyphion . . .	C.	157	4	2	
X Y Z . . .	G.	128	2	2	
Yap . . .	C.	91	1	4	
Yarmsdale . .	M.	355	4	A.	
Yell . . .	F.	40	4	2	
Yellowbird . .	G.	254	2	4	Bird, 206 = 8.
Yellow Danger .	C.	304	7	4	Danger, 258 = 6.
Yellow Flower . .	M.	364	4	A.	Flower, 318 = 3.
Yellow Plush . .	G.	456	6	6	Plush, 410 = 5.
Yellow Vixen . .	M.	180	9	A.	Vixen, 136 = 1.
Yemassee . .	C.	121	4	4	
Yenikale . .	M.	120	3	5	
Yeaman II. . .	G.	110	2	5	
Yorick . . .	G.	232	7	4	
Yorktown . . .	G.	688	4	4	
You Go Off . .	F.	124	7	2	
Young Courtenay .	G.	766	1	4	Courtenay, 686 = 2.
Young Lochinvar .	G.	463	4	4	Lochinvar, 383 = 5.
Young Neville . .	H.	250	7	5	Neville, 170 = 8.
Ypsilanti . .	C.	632	2	4	
Yumboe . . .	G.	58	4	5	
Yusen . . .	H.	130	4	6	
Zagiga . . .	M.	49	4	5	
Zam . . .	C.	48	3	2	
Zanita II. . .	M.	459	9	A.	
Zanoni . . .	G.	119	2	A.	
Zarinthia . .	F.	674	8	3	
Zatty . . .	F.	418	4	3	
Zaza . . .	F.	16	7	2	
Zaza II. . .	M.	16	7	6	
Zerline . . .	M.	297	9	5	

Name.	Sex.	Value.	Remainder.	Age (figures indicate years and "A" aged).	Remarks.
Zest	F.	477	9	3	
Zig-a-Zag . . .	G.	56	2	A.	
Zimbro . . .	G.	255	3	5	
Zimra . . .	F.	248	5	4	
Zinfandel . . .	C.	232	7	2	
Zinga . . .	F.	78	6	3	
Zither . . .	F.	612	9	4	
Zodiac II. . .	G.	48	3	A.	

NOTE.—(1) In case of plural names, the remainders in the remarks column will be found more reliable ; but for the purposes of comparing the full values the sums in the value column should be used.

(2) Little reliance should be placed in case of the numbers of the unnamed ones, whose values are derived from the names of their dams.

(3) In case of new names, 7 months should elapse before they can be used with confidence ; until then it is better to take the names of the dams instead.

(4) When a selection is to be made from a company of similar "remainders" of equal ages, it is much better to leave it alone.

(5) Single names are most powerful and thoroughly reliable, especially when the ages of the competitors are different.

(6) Foreign names should be very carefully dealt with, as the least mistakes in pronunciation might affect the results.

APPENDIX B.

Glaring Instances of the Names beginning with the Sounds belonging to the same Planets being placed 1st, 2nd, and 3rd, or 1st and 2nd.

Maiden Erlegh, 26–11–1901—

    4th race . . . Marriage Lines.
                             Tarolinta.
                             Molester.

Leicester, 27–11–1901—

    1st race . . . Nelson.
                             Chair of Kildare.
                             Laplander.

Kempton Park, 29–11–1901—

    3rd race . . . Overrated.
                             Oban.
                             Excepcional.

Kempton Park, 30–11–1901—

    1st race . . . Tonsure.
                             Traveller.

Gatwick, 4–12–1901—

    5th race . . . Bell Sound.
                             Vincent.

Wye, 5–12–1901—

    1st race . . . Kurvenal.
                             Peopleton.

    2nd race . . . Snapshot.
                             Shifter.

Shirley, 9–12–1901—

    1st race . . . Postman's Knock.
                             Plumage.

MANCHESTER, 10–12–1901—

    4th race . . . Loddon.
    Arnold.

    6th race . . . Stirtloe.
    Glory Hole.
    Greek Lass.

PLUMPTON, 11–12–1901—

    2nd race . . . Celer.
    Goldwasher.

PLUMPTON, 12–12–1901—

    1st race . . . Blairgowrie.
    Revera.

    5th race . . . Olive Branch.
    Livorno.

KEMPTON PARK, 26–12–1901—

    5th race . . . Key West.
    Peruke.
    Postman's Knock.

KEMPTON PARK, 27–12–1901—

    1st race . . . Merry Monk.
    Master Herbert.

    2nd race . . . Goosey Gander.
    Sparsholt.

    5th race . . . Netherland.
    Lord James.

LIMERICK, 26–12–1901—

    5th race . . . Green Witch.
    Gadwall.

LEOPARDSTOWN, 26–12–1901—

    4th race . . . Lady Flight.
    Ledessan.

HURST PARK, 28–12–1901—

    5th race . . . Snowden.
    Slingsby.

KEELE PARK, 31–12–1901—

    2nd race . . . Glen Royal.
    Speculation.

    5th race . . . Nelson.
    Loughran.

BALDOYLE (METROPOLITAN), 1–1–1902—

    2nd race   .   .   .   Johnny Mack.
                                       Sallypark.

MANCHESTER, 1–1–1902—

    1st race   .   .   .   Trouvere.
                                       Metheolis.

WINDSOR, 8–1–1902—

    3rd race   .   .   .   Alone in London.
                                       Jap.

PLUMPTON, 11–1–1902—

    1st race   .   .   .   Wiki Wiki.
                                       Blisworth.

BIRMINGHAM, 13–1–1902—

    2nd race   .   .   .   Pirate's Bride.
                                       Prince George.

MANCHESTER, 15–1–1902—

    1st race   .   .   .   Coolock.
                                       Kitchener.

MANCHESTER, 16–1–1902—

    3rd race   .   .   .   Caerleon.
                                       King David.

    5th race   .   .   .   Well Fort.
                                       Bevil.

HURST PARK, 17–1–1902—

    2nd race   .   .   .   Spinning Boy.
                                       Shackleford.
                                       Goldwasher.

LINGFIELD, 22–1–1902—

    4th race   .   .   .   Perdicus.
                                       Pomfret.

    6th race   .   .   .   M'Mahon.
                                       Mercury.

LINGFIELD, 23–1–1902—

    2nd race   .   .   .   Slemish.
                                       Speculation.

KEMPTON PARK, 24–1–1902—

    5th race   .   .   .   Tonsure.
                                       Tin Soldier.

WINDSOR, 27–1–1902—
   2nd race . . . Jove.
            Greenhall.

GATWICK, 29–1–1902—
   5th race . . . Menelik.
            Morville.

   6th race . . . Glen Choran.
            Shannon Lass.

GATWICK, 30–1–1902—
   3rd race . . . Bevil.
            Barsac.

SANDOWN PARK, 1–2–1902—
   1st race . . . College Queen.
            Hedera.
            Curlew.

FOLKESTONE, 3–2–1902—
   3rd race . . . Mystic Moon.
            Mayfly.

FOLKESTONE, 4–2–1902—
   4th race . . . Coroun.
            Prince Leo.

LEICESTER, 5–2–1902—
   3rd race . . . Shepherd King.
            Golden Rule.

BIRMINGHAM, 25–2–1902—
   1st race . . . Checkman.
            Chocolate.
            Love Child.

PLUMPTON, 25–2–1902—
   5th race . . . George Fordham.
            Senateur.

KEMPTON PARK, 26–2–1902—
   5th race . . . Bramante.
            Bonnie Yorkshire Lad.

HURST PARK, 1–3–1902—
   5th race . . . Venetian Monk.
            Venerable Bede.

SOUTHWELL, 4–3–1902—

     1st race . . . Tyna.
                       Tankerness.

     4th race . . . Ortygian.
                       Naomi.
                       Alpheus.

GATWICK, 5–3–1902—

     6th race . . . Miss Grab.
                       Muggins.

N.B.—The reader may himself examine the annual records of horse-racing, and he is sure to find abundant evidence of the law of sound being in operation. The above are only a very few instances, just for the purposes of illustration.

PRINTED BY MORRISON AND GIBB LIMITED, EDINBURGH.

CPSIA information can be obtained
at www.ICGtesting.com
Printed in the USA
BVHW080922040419

544590BV00002B/23/P